Lars Hegenbart

Numerical Efficiency Calibration of In Vivo Measurement Systems

Monte Carlo Simulations of In Vivo Measurement Scenarios for the Detection of Incorporated Radionuclides, Including Validation, Analysis of Efficiency-Sensitive Parameters and Customized Anthropomorphic Voxel Models

Numerical Efficiency Calibration of In Vivo Measurement Systems

Monte Carlo Simulations of In Vivo Measurement Scenarios for the Detection of Incorporated Radionuclides, Including Validation, Analysis of Efficiency-Sensitive Parameters and Customized Anthropomorphic Voxel Models

by
Lars Hegenbart

Dissertation, Universität Karlsruhe (TH)
Fakultät für Elektrotechnik und Informationstechnik,
Tag der mündliche Prüfung: 27. Juli 2009
Referent: Prof. Dr. Ing. Manfred Urban
Korreferent: Prof. Dr. Ing. Uwe Kiencke

Impressum

Karlsruher Institut für Technologie (KIT)
KIT Scientific Publishing
Straße am Forum 2
D-76131 Karlsruhe
www.ksp.kit.edu

KIT – Universität des Landes Baden-Württemberg und nationales
Forschungszentrum in der Helmholtz-Gemeinschaft

KIT Scientific Publishing 2010
Print on Demand

ISBN 978-3-86644-509-3

Zusammenfassung

Effizienzkalibrierungen der Detektoren von Inkorporationsmesssystemen beruhen auf dem Einsatz von physikalischen Phantomen, die mit radioaktiven Quellen bestückt werden können um verdächtigte Inkorporationsfälle nachzuahmen. Systematische Fehler der herkömmlichen Kalibrierverfahren können beträchtliche Über- oder Unterschätzungen der inkorporierten Aktivität und damit auch der absorbierten Dosis verursachen.

In dieser Arbeit werden Monte Carlo Methoden angewendet um Strahlentransportprobleme zu lösen. Virtuelle Modelle der Messeinrichtungen des In-vivo Messlabor des Instituts für Strahlenforschung, inklusive der Detektoren und anthropomorpher Phantome, wurden entwickelt. Software-Werkzeuge wurden entwickelt um den Umgang mit speicherintensiven anthropomorphen Modellen zu erleichtern bei der Visualisierung, sowie die Vorbereitung und Auswertung von Simulationen von In-vivo Messsituationen mit solchen Modellen. Die Simulationsumgebung mit den benutzten Werkzeugen, Methoden und Modellen wurde validiert.

Die Sensitivität verschiedener Parameter auf die Detektoreffizienz wurde untersucht um mögliche systematische Fehler zu identifizieren und auch zu quantifizieren. Massnahmen auch im Hinblick auf deren Anwendung in der Routine wurden ergriffen um die Bestimmung der Detektoreffizienz zu verbessern.

Ein Positionsmesssystem wurde entwickelt und in der Teilkörperzählkammer installiert um die genaue Position und die Abstände von Detektoren relativ zu den Probanden zu bestimmen, die als sensitive Parameter identifiziert wurden. Rechner wurden zu einem Verbund zusammengeschaltet um Monte Carlo Simulationen zu vereinfachen und Rechenzeiten zu verkürzen. Bildregistrierungstechniken wurden entwickelt um vorhandene anthropomorphe Modelle zu transformieren anhand der Anatomie von Probanden. Die entwickelten Massnahmen und Methoden haben die bisher angewendeten klassischen Methoden erfolgreich verbessert.

Abstract

Detector efficiency calibration of in vivo bioassay measurements is based on physical anthropomorphic phantoms that can be loaded with radionuclides of the suspected incorporation. Systematic errors of traditional calibration methods can cause considerable over- or underestimation of the incorporated activity and hence the absorbed dose in the human body.

In this work Monte Carlo methods for radiation transport problem are used. Virtual models of the in vivo measurement equipment used at the Institute of Radiation Research, including detectors and anthropomorphic phantoms have been developed. Software tools have been coded to handle memory intensive human models for the visualization, preparation and evaluation of simulations of in vivo measurement scenarios. The used tools, methods, and models have been validated.

Various parameters have been investigated for their sensitivity on the detector efficiency to identify and quantify possible systematic errors. Measures have been implemented to improve the determination of the detector efficiency in regard to apply them in the routine of the in vivo measurement laboratory of the institute.

A positioning system has been designed and installed in the Partial Body Counter measurement chamber to measure the relative position of the detector to the test person, which has been identified to be a sensitive parameter. A computer cluster has been set up to facilitate the Monte Carlo simulations and reduce computing time. Methods based on image registration techniques have been developed to transform existing human models to match with an individual test person. The measures and methods developed have improved the classic detector efficiency methods successfully.

Acknowledgements

First of all I would like to thank Professor Dr. Manfred Urban for supporting my way through the qualification phase and for being a helpful advisor at all times. I extend my appreciations also to Professor Dr. Uwe Kiencke for reviewing this thesis as the second examiner.

I would to like to thank Professor Dr. X. George Xu and his team of post-docs and doctoral candidates (Dr. Valery Taranenko, Dr. Binquan Zhang, Matt Mille, Dr. Yong Hum Na, and Dr. Juying Zhang) of Rensselaer Polytechnic Institute for providing me the RPI Adult Male and Female voxel models and for the fruitful team-work on the breast-size project during the academic exchange in spring 2008.

This exchange was supported by the Karlsruhe House of Young Scientists (KHYS). Here, I would like to thank the KHYS-team Dr. Britta Trautwein and Gaby Weick for their help.

I also would like to thank Professor Dr. Olaf Dössel and Dr. Gunnar Seemann and their team at the Institute of Biomedical Engineering for support on the MEET Man.

I extend my appreciations to Professor Dr. Jürgen Lehmann and colleagues of the Vincentius Kliniken Karlsruhe for scanning the torso phantom.

Thanks for professional help and advice goes to my supervisors and colleagues Andreas Benzler, Dr. Bastian Breustedt, Dr. Bernd Heide, Dipl.-Ing. Wolfgang Klein, Dr. Debora Leone, Ing. Olaf Marzocchi, and Dr. Jutta Schimmelpfeng.

As well, I would like to thank Dipl.-Ing. (BA) Steffen Sessler and all other students including Corey Prumo for working with me part-time on the project during their diploma projects or internships.

I would like to thank Dipl.-Ing. Siegfried Ugi and Dipl.-Ing. Bernd Reinhardt and their teams for the mechanical and electrical installation of the positioning system and for the design and construction of the frame for the HPGe-detector. I extend my appreciations also to the Body Counter Team, as well as

Dipl.-Ing. Christoph Wilhelm and Dr. Max Pimpl including their teams for their help with measurements and assistance.

I would like to thank Bärbel Stuck and all other colleagues and friends of Hauptabteilung Sicherheit and the new founded Institute of Radiation Research who are not mentioned explicitly for a good collaboration and colleagueship.

Thanks also to Solveig, Marion, and Marey for proofreading this work. Finally I would like to thank Fleur for supporting my work with encouragement.

Contents

Chapter 1

Introduction

1.1 In Vivo Measurement at KIT

The Institute of Radiation Research (ISF[1]) at the Karlsruhe Institute of Technology (KIT) operates an in vivo measurement laboratory (IVM[2]). The task of the IVM is to perform in-vivo bioassay measurements periodically on workers who encounter open radioactive material. Measurements are also performed in the event of a suspected incident. Such incidents are ingestion, inhalation, or permeation through wounds of radionuclides by accident.

The task of the IVM is to determine the activity of the radionuclide by measurements with detectors from the outside of the body. The detectors count a fraction of photons emitted from radioactive nuclides within the human body. The detectors used at IVM are described in section 2.1. The activity can be calculated from the count rate when the detector efficiency (equation 1.1) is known. Once the activity is determined the absorbed dose that the worker received can be estimated. Then, medical treatments can be adjusted appropriately.

---

[1]German: Institut für Strahlenforschung
[2]German: In vivo Messlabor

$$cps = \eta \cdot A \tag{1.1}$$

with

$$cps = \text{count rate (counts/s)}$$
$$\eta = \text{detector efficiency (counts/decay)}$$
$$A = \text{activity (decays/s = Bq)}$$

In the human body two basic types of deposition of radionuclides can be found. In the first type, the deposition is distributed uniformly throughout the entire body. In the second type only specific tissues or organs of the body are affected. The elapsed time after the uptake of the radionuclide may have an influence on its distribution in the body. The radionuclide's metabolization is determined by its chemistry and its biokinetic behavior and affects in which organ it is accumulated or if it is excreted rapidly.

At the IVM two measurement chambers are available. In the Whole Body Counter (WBC) uniform depositions are investigated, while in the other chamber, the so called Partial Body Counter (PBC), activities of single organs (e.g. lung, thyroid, liver, bone, ...) or wounds are measured. The WBC and the PBC at IVM are described in details in section 2.1.

1.2   Challenges of Detector Efficiency Calibration

Each type of measurement needs appropriate calibration methods. The precision of the determination of the activity in WBC and PBC depends substantially on systematic errors of the detector efficiency calibration. Detector efficiency is given by the ratio of counted photons in a detector and decays of a radionuclide (equation 1.1). The term detector efficiency is explained in detail in section 3.1. The calibration of the WBC and the PBC is routinely performed by measurements of a phantom. It consists of material which approximates the density and the effective atomic number of human tissue, to provide equivalent radiation attenuation as the body of the person.

Calibration errors occur especially in low-energy photon emitters such as ^{210}Pb (46.5 keV), ^{234}Th (63.3 keV), ^{239}Pu (X-rays from 13 to 20 keV), or ^{241}Am (59.5 keV) and also β-emitting radionuclides like ^{90}Sr, that are detected by Bremsstrahlung. Known reasons for the systematic errors, amongst others, are the following:

- *Inhomogeneous distribution* of the radionuclide in the body in different retaining organs, or also inhomogeneities within one organ.

- *Detector position* relative to the source of the radiation, which might be different to standard position, for instance those used for calibration.

- *Scattering effects* caused by the measurement equipment, the walls, and the test person itself depending on its anatomy.

- *High absorption* of low-energy photon emitters in overlying tissue. The tissue composition and thickness may vary locally and cannot be determined accurately.

Lung counting, i.e. measurements of activities in the lung, is an example for a challenging detector efficiency calibration. Radionuclides of interest are usually low-energy photon emitters that are attenuated by the chest wall of the test person. Lung counting is performed with detectors on the front side of the chest. While the lung itself has a low density, the overlying tissue has not. The thickness of the chest wall and its adipose content are of importance. The distribution of activity in the lung depends on radioactive particle size, breathing rate, and personal habits such as smoking. Depending on the time since the inhalation and the radionuclide compound's solubility and biochemistry, the activity may further distribute through the body. Kramer et al. (1999; 2000b) determined large uncertainties of the lung activity estimates due to the distribution in a calibration phantom compared to the one in the test person (Kramer and Hauck, 1999; Kramer *et al.*, 2000b).

Several phantoms, representing the human body, are used at the IVM (for details see section 3.2) for detector efficiency calibration. The phantoms are standardized to represent an average human body. The distribution of radioactive sources, which can be loaded into the phantom, is predefined by the phantom manufacturer. Overlays are used to alter the chest wall thickness in discrete steps. A more flexible usage for instance altering more anthropometric parameters of such phantoms is not possible. The practical experience shows that determined calibration efficiencies can not necessarily be transferred to a specific case. This leads to a considerable uncertainty in the determination of the activity and hence the estimation of the absorbed dose.

1.3   State of the Art - Numerical Efficiency Calibration

Since computers became faster and the amount of memory higher, Monte Carlo methods (see section 4.1) are used more often by scientists to determine detector efficiencies numerically. Monte Carlo methods are computational algorithms that are based on repeated random sampling with a pseudo-random number generator. They are often used to simulate physical systems, which are infeasible or impossible to determine with an analytical algorithm. In this investigation a code is used that simulates the transportation of photons, electrons and other particles through a customizable virtual world.

In order to use such codes, the WBC and PBC at the IVM need to be modeled correctly. Although it is not completely trivial, but all objects, for instance detectors, examination tables, and walls of the chamber can be modeled easily in this virtual world. A main challenge is to model individual people. Human models (see section 4.4) can be used in the virtual world as calibration phantoms. Such human models can be created from computer- or magnetic resonance tomography scans. These models are made of small elements called *voxels*. The term voxel is derived from volumetric a picture element (pixel). Voxel models are not only used in medicine. For instance, they are used to calculate electric and magnetic fields in the body. An example for this is the MEET Man (Sachse *et al.*, 2000), which was created at the Institute for Biomedical Engineering (IBT[3]) from image data from the Visible Human Project (Ackerman,

---

[3]German: Institut für biomedizinische Technik

1991). The ISF has a copy of the MEET Man. The first steps for this work have been performed with this voxel model.

Once a model of a complete in vivo measurement scenario is set up and valid, it can be used as a tool for numerical efficiency calibration and then detector efficiencies can be calculated by Monte Carlo simulations. The simulation of the WBC and the PBC offers various possibilities. Virtual experiments can be performed with Monte Carlo simulations, which might never be performed in the real laboratory. It is expected that individual models improve the calibration method and reduces the systematic errors.

1.4   Objective and Structure of this Work

The aim of this work is to generate a valid environment for Monte Carlo simulations of in vivo measurement scenarios. The focus lies on human models and how detector efficiencies are influenced by the model and other parameters. Human models are coded with numbers and mathematical equations and hence can be modified. On the contrary to physical phantoms, customizing a computer model of a human body is a non-destructive method.

The final objective of this work is to create new models that match an individual test person better than the conventional physical phantoms. Therefore the possibilities of creation and application of individual voxel models are examined for numerical efficiency calibration at the IVM.

Chapter 2 describes the equipment used and the basics for detecting photons from within the human body. Classical efficiency calibration is introduced in chapter 3. It was necessary to create a whole simulation environment, which is described in section 4. There, also the developed tools for this work are introduced and the creation of a voxel model of an existing physical phantom is presented. The simulation environment needed to be validated. Chapter 5 deals with this validation. Once a valid simulation environment was achieved, it could be used for sensitivity analyses. The influence of various parameters on the detector efficiency have been checked in chapter 6. After the sensitivity of each parameter was investigated, methods have been developed that create individual models. Chapter 7 discusses this in detail. In chapter 8 the results and the outcome of this work are summarized and a perspective for further research is pointed out.

Chapter 2

Detection of Photons

Every radionuclide has characteristic emissions. When the decay of a radionuclide takes place inside the human body, α- particles are completely absorbed. β-particles are strongly attenuated as well within the body, while photons – depending on their energy – are able to reach the outside of the body. Photon attenuation is described in the appendix A.2. The emitted photon energies are characteristic for each radionuclide.

Scintillation or semiconductor detectors are commonly used to measure photons so that the activity (see appendix A.1) and the resulting absorbed dose from the radionuclide in the body can be estimated.

2.1 Measurement Equipment

The measurement takes place in a shielded chamber to reduce background radiation resulting from cosmic radiation, radioactive nuclides in building materials, radon gas from subsurface sources, and its mostly airborne decay products. The reduction of the background lowers the detection limit of the detector system and makes detection of small activities possible.

There are two measurement chambers at the IVM. One is designed for whole body counting and the other is designed for partial body counting. The **whole body counter** (WBC) has four cylindrical NaI(Tl)[1]-detectors

---

[1]Thallium doped sodium iodide

with a crystal diameter of 20.32 cm and 10.16 cm thickness. Test persons
are measured in a supine position on an examination table in chamber
with a labyrinth entrance. The chamber is shielded with walls of 15 cm
thick steel (25 cm at the floor) and 0.96 cm lead. Two detectors are
fixed above, and two are below the examination table to cover the whole
body (Figure 2.1). The detector crystals are each connected to a photo-
multiplier, a pre-amplifier, and multichannel analyzer (MCA) with 256
channels. The photon energy range covered by the system is from
approximately 100 to 2500 MeV. The MCA is read out by Canberra's
Genie 2000 software. An in-house developed software is used to evaluate
the spectra.

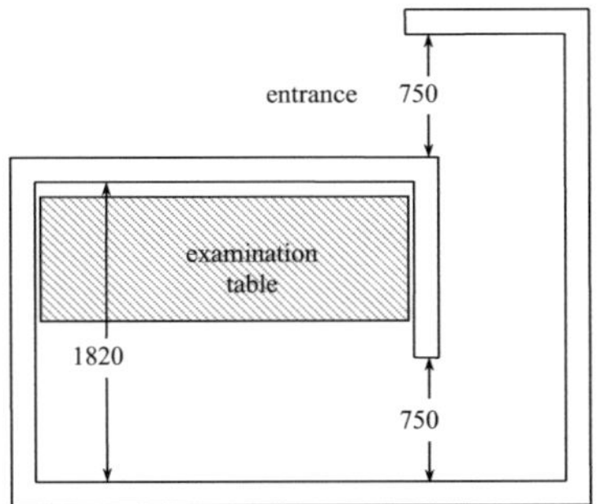

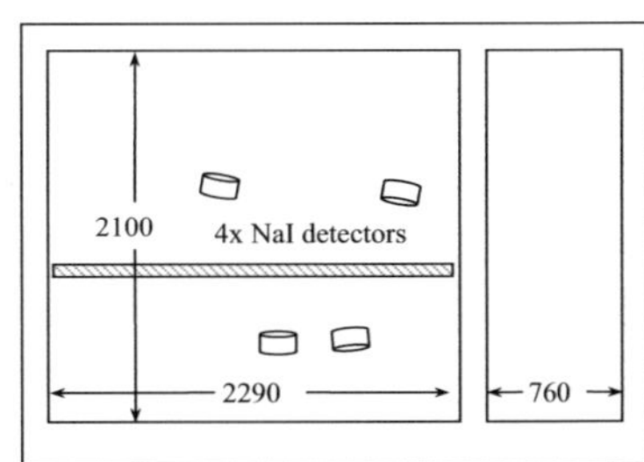

Figure 2.1: Sketch from the WBC chamber at IVM (Sessler, 2007). The
left drawing shows the chamber from bird's eye view. The right drawing
shows a cross-section through the eastern and western wall. The positions
of the four detectors are depicted by cylindrical objects. The distances are
given in millimeters.

Figure 2.2 shows the chamber of the **partial body counter** (PBC). It
is also shielded by walls of 15 cm thick steel. The floor is made of 25 cm
thick steel. In addition, the chamber inside has a graded-Z shielding[2] that
consists of layers of 5 mm lead, followed by 1.5 mm tin, and innermost
0.5 mm of copper. The door is motor-driven due to its heavy weight. The
test persons lie in a supine position on examination table in the shielded
chamber.

In the PBC, most commonly two out of three phoswich detectors

---

[2]Z refers to the atomic number, i.e. the number of an atom's protons

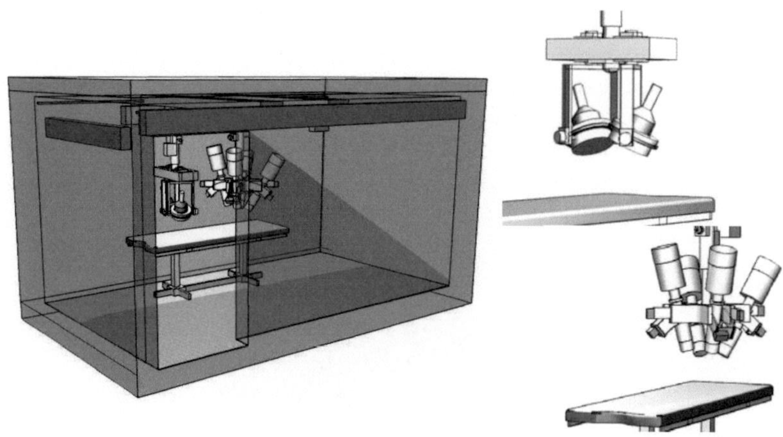

Figure 2.2: Sketches from the PBC chamber at IVM generated by data from a laser scanner. The left drawing gives an three-dimensional, semi-transparent insight into chamber with the couch and two detector systems. The door (in front) was removed. The upper right drawings show the phoswich detector pair and its rack. The lower right drawing represents a liquid-N-cooled HPGe-detector system, which was not considered in this thesis.

from the 1970s are used for detecting low-energy photon emitters. Two of the three detectors (#1 and #2) are mainly used for lung and liver measurements (Figure 2.3), but also for knee, skull, or lymph nodes measurements. The phoswich detector pair is mounted on a movable rack and each detector can be adjusted widely in its inclination and position. Section 6.3.1 gives more details about the detector rack later in this work. The third phoswich detector is used for measurements at the skull and can be moved below the head of the subject. All phoswich detectors are composed of a combination of two cylindrical scintillation crystals with a diameter of 20.32 cm positioned axially aligned onto each other. A 1-mm beryllium layer covers the entry window of the detector. It is followed by a light reflecting powder (aluminum oxide) layer. The first scintillation crystal is 1-mm thick NaI(Tl) followed by the second 5.08 cm thick CsI(Tl)-crystal. After the the CsI-crystal a third crystal follows. It is made of undoped NaI and serves for light conduction to the photo-multiplier tube. The crystals are surrounded by Al_2O_3 powder and mounted in a steel case.

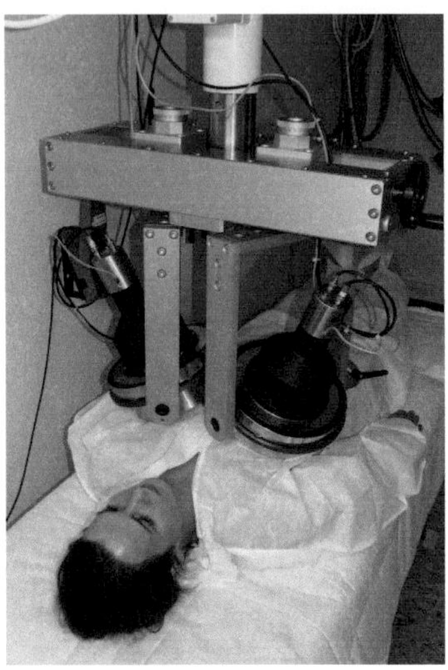

Figure 2.3: Rack holding the two phoswich detectors in the partial body counting chamber. The test person lies on the bed in supine position. The detectors are positioned for a lung measurement.

The photo-multipliers are connected to pre-amplifier and MCAs. Canberra's Genie 2000 software reads out the 256 channels of the MCAs. The special property of the phoswich detector system is a pulse-shape analyzer. It can distinguish between pulses from the NaI(Tl)- and those from the CsI(Tl)-crystal. Different time constants (risetimes) of the pulses originating from the different crystals result in electronically distinguishable pulse shapes. This way, counts from the NaI(Tl)- and the CsI(Tl)-crystal can be collected separately. The thin NaI(Tl)-crystal absorbs most of the low-energy photons. The channels of this part of the detector are adjusted to cover the energy range from roughly 10 to 250 keV. The channels for the CsI(Tl)-crystal cover the energy range from approximately 10 to 2500 keV.

The pulse-shape analyzer works on the principle that time constants of the excited state of the scintillation process in the two crystals are different. This results in different pulse shapes. An anti-coincidence circuit discards detector signals, if both pulse shapes are analyzed at the same time. This technique reduces the contribution from Compton scattering and background radiation in the low-energy range of the spectrum.

The background radiation is measured once a week over night in both chambers. For activity calculations, the background spectrum is always normalized and subtracted from the recorded spectrum. Unfortunately the *background* originating from the subject itself cannot be reduced. Primordial ^{40}K makes up 0.012% of the natural occurring potassium, which is an essential mineral in the human body. In all spectra the gamma line at 1461 keV is visible. In addition, ^{40}K is a β-emitter, so it produces a broad continuous low-energy spectrum caused by Bremsstrahlung photons.

At the moment a postgraduate is doing research with four new Canberra XtRa **HPGe-detectors**[3], which are intended to replace the NaI(Tl)-detectors in the WBC. Due to a thin detector entrance window and a thin dead layer, the energy range goes down to 10 keV. Therefore, they can be used also for detecting actinides in partial body counting. The energy resolution is with a few keV much better, but the detector efficiency is low compared to the NaI(Tl)-detectors due to the small size of the germanium crystals. 8192 channels cover the energy range from 10 to 2048 keV. The new detectors are cooled electrically without liquid nitrogen. Detailed technical drawings are available.

2.2 Energy Calibration

Energy calibration of a detector system is needed to identify the radionuclides in a recorded energy spectrum. This calibration is usually done with a number of known radionuclides with unambiguous single X-rays or gamma-lines throughout the whole energy range. Figure 2.4 shows a typical spectrum of an ^{241}Am point source recorded by a NaI(Tl)-crystal of a phoswich detector. Often, due to the limited energy resolution of NaI(Tl)-detectors single lines cannot be resolved. Table 2.1 shows the characteristic photon energies of ^{241}Am and their emission probability per decay.

---

[3]High-Purity Germanium

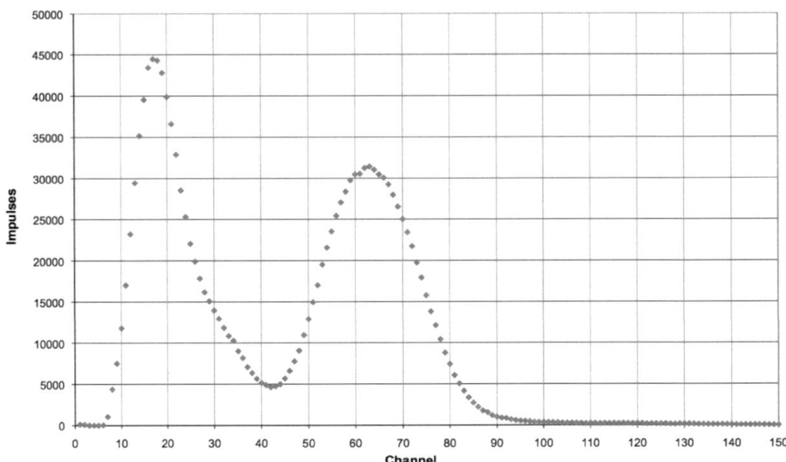

Figure 2.4: ^{241}Am point source spectrum recorded by a NaI(Tl)-crystal detector of a phoswich system.

For an energy calibration, one needs to assign the channels of a detector system to correct energy values. To do this, a calibration source is placed in front of the detector without other objects nearby, which may cause scattering. Scattering can cause a shift of the peak to lower energy. In the example (figure 2.4) the peak center of the right peak results from the 59.5-keV-line of ^{241}Am and channel 63 can be assigned to that energy. Usually, there is a linear relationship between energy and channel number. Nevertheless, in rare cases there are deviations from the linearity of the amplifier-analyzer system (Debertin and Helmer, 1988). The Genie 2000 software offers the user an additional correction term and describes the energy-channel-relationship by a polynomial of second order (equation 2.1). For example, the parameter w was calculated for HPGe-detectors. Low values roughly in the range 10^{-7} to 10^{-9} keV^{-2} have been observed. The example shown in figure 2.5 illustrates that $w \ll v^2$. Therefore, it was neglected to use the w parameter for all measurements performed in this work. The Genie 2000 software does the channel/energy assignment automatically (figure 2.5) by assigning a couple of peak centers to radionuclides stored with their peak data in an electronic library. The more peaks used for the calibration, the more

Table 2.1: ^{241}Am emission probabilities (Schötzig and Schrader, 2000).

Energy (MeV)	Emission probability (decay^{-1})	Energy (MeV)	Emission probability (decay^{-1})
0.012	8.37×10^{-03}	0.070	2.90×10^{-05}
0.014	1.08×10^{-02}	0.076	5.90×10^{-06}
0.014	1.19×10^{-01}	0.099	2.03×10^{-04}
0.016	3.77×10^{-03}	0.103	1.95×10^{-04}
0.018	1.86×10^{-01}	0.123	1.00×10^{-05}
0.021	4.82×10^{-02}	0.125	4.08×10^{-05}
0.026	2.40×10^{-02}	0.147	4.61×10^{-06}
0.032	1.74×10^{-04}	0.170	1.73×10^{-06}
0.033	1.26×10^{-03}	0.208	7.91×10^{-06}
0.043	5.50×10^{-05}	0.323	1.52×10^{-06}
0.043	7.30×10^{-04}	0.332	1.49×10^{-06}
0.056	1.81×10^{-04}	0.335	4.96×10^{-06}
0.058	5.20×10^{-05}	0.367	7.90×10^{-06}
0.060	3.59×10^{-01}	0.369	2.17×10^{-06}
0.065	1.45×10^{-06}	0.377	1.38×10^{-06}
0.067	4.20×10^{-06}	0.662	3.64×10^{-06}
		0.722	1.96×10^{-06}

accurate the software can fit the energy calibration function.

$$S(E) = u + vE + wE^2 \tag{2.1}$$

with

$S(E)$ = channel calibration function; gives the channel number

E = energy in (keV)

u = parameter

v = parameter in (keV^{-1})

w = parameter in (keV^{-2})

According to the most recent standard operating procedure (SOP) (Mohr and Cordes, 2008) the energy-channel relationship of the NaI(Tl) detector of the phoswich system is calibrated daily, manually by checking the peak maxima of an ^{241}Am source. If needed, the gain of the detectors'

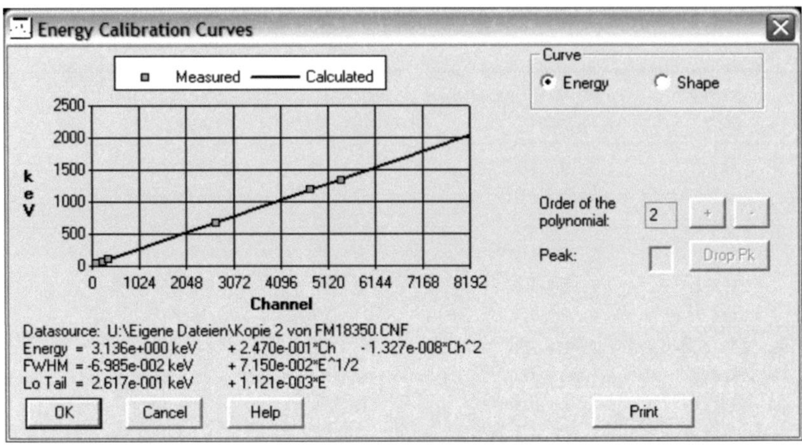

Figure 2.5: Screenshot from Genie 2000 shows a calibration curve of HPGe-detector. The determined parameter for the quadratic term is low. The curve is almost linear.

amplifiers is adjusted. The 17-keV- and the 59.5-keV-peak of ^{241}Am are used for energy calibration. The maximum of the peaks is adjusted to be in channel 17 and channel 61, respectively.

2.3 Energy Resolution

The energy resolution of a detector is defined with the full width at half maximum (FWHM) of the peak of a specific energy. In theory, the peak is a sharp line but due to scattering, scintillation effects, and electronic effects, the sharp line is broadened and can be approximated with a Gaussian curve. The energy resolution can be described as a function of the energy (equation 2.2).

$$FWHM = a + b\sqrt{E} + cE^2 \tag{2.2}$$

with

$$FWHM = \text{detector resolution in (MeV)}$$
$$E = \text{energy in (MeV)}$$
$$a = \text{parameter in (MeV)}$$
$$b = \text{parameter in (MeV}^{1/2})$$
$$c = \text{parameter in (MeV}^{-1})$$

for Gaussian approximation

$$FWHM = 2\sqrt{2\ln 2}\,\sigma \tag{2.3}$$

with

$$\sigma = \text{standard deviation of normal distribution}$$

Chapter 3

Classical Efficiency Calibration

This chapter describes the classical efficiency calibration of the in-vivo measurement systems at IVM. It introduces the basics about efficiency calibration and the physical calibration phantoms. Efficiency calibration is necessary to calculate the activity of supposed radionuclides in the body of a test person.

3.1 Detector Efficiency

First the term detector efficiency needs to be defined. Generally, detector efficiency η_d is the quotient of measured counts in the detector and emitted photons from the radionuclide. Equation 3.1 extends equation 1.1 introduced in chapter 1.

$$\eta_d = \frac{N}{A(\tau) \cdot t} \tag{3.1}$$

with

η_d = detector efficiency in (counts/decay = counts s^{-1} Bq^{-1})

N = number of counts in (counts)

$A(\tau)$ = activity during the measurement in (decays/s = Bq)

t = life-time of detector during measurement in (s)

Usually a range of channels are summed up to give the total counts of a peak in the spectrum. For each radionuclide, there are energy ranges defined as *regions of interest* (ROI). The idea of the ROI is that the measurement only focusses on the counts from one radionuclide. Therefore, other regions in the spectrum are *not interesting* for evaluation of the activity of the radionuclide. A standard value (BUM, 2007) for the ROI is ± 1.25 FWHM around the theoretical energy of the peak. Thus it is often called peak area.

Care must be taken in case more than one photon-emitting radionuclide is present. Peaks increase the counts in every channel in the spectrum below, due to scattering effects. If the peak of interest has a lower energy than a second peak, which is caused by a second radionuclide, the peak area will be larger. This error can be corrected by subtracting the base counts, sometimes refered as the *continuum*, the peak of interest sits on.

The ROI differs not only for each radionuclide, it also varies for each detector due to a different energy resolution. The ROI is broad for scintillation detectors like NaI(Tl)-crystals due to its limited energy resolution. For semi-conductor detectors like HPGe-detectors, the ROI can be chosen relatively narrow around the peak.

Equation 3.2 is actually used in practice for detector efficiency and extends equation 3.1 by the ROI. With known detector efficiency in a ROI for a specific radionuclide and the detector life-time[1], one can calculate

---

[1] the time, the detector system is *alive* and registers counts, due to the duration of the signal processing; the life-time is usually shorter than the real time

the activity (see also appendix A.1) and then estimate the absorbed dose.

$$\eta_{d,l,u} = \frac{\sum_{i=l}^{u} N_i}{A(\tau) \cdot t} \tag{3.2}$$

with

$\eta_{d,l,u}$ = detector efficiency in ROI [l,u] in (counts/decay)

N_i = number of counts in channel i in (counts)

l = lower channel number of ROI

u = upper channel number of ROI

With large single detectors or detector arrays, one tries to maximize the surface and hence the probability that a photon enters the detector. In partial body counting the detectors are placed on or near the surface of the body to minimize the distance to the target organ. This maximizes the counting efficiency. A high detector efficiency has the advantage of short measurement times for the test person. A high number of counts results in a low statistical counting error, and the measurement signal emerges clearly from the background radiation.

In order to get results with low statistical counting errors, a high number of counts in the ROI is required. The spectrum can be checked visually for its quality. Such errors are low, if the spectrum's curve is smooth. Noise will make the curve look rough. This is an evidence that the counting error might be high. According to the central limit theorem, the counting error is proportional to the reciprocal square root of the number of counts. Thus the counting error can be quantified with the relationship shown in equation 3.3.

$$\sigma \propto \frac{1}{\sqrt{N}} \tag{3.3}$$

with

σ = relative counting error

N = number of counts

3.2 Physical Phantoms

The determination of the detector efficiency is more complicated than the energy calibration. Typically, detector efficiency calibrations are performed with physical phantoms that are spiked with known radionuclides and radioactivities in the organ of interest. Ideally, the phantom is identical or at least similar in anatomy to the measured person. An overview about calibration phantoms is given in an ICRU-Report (ICRU, 1992).

The IVM has a few calibration phantoms mainly used in the PBC. They are made of tissue-equivalent material and that can be loaded with a variety of defined radioactive sources.

- The Lawrence Livermore Realistic Torso Phantom (or short *LLNL Torso Phantom*) is an anthropomorphic torso phantom with exchangeable organs for the simulation of homogenous nuclide depositions in lung and liver (Griffith *et al.*, 1978, 1986). Radioactive lungs and livers are loaded with either one of the following nuclides or nuclide mixtures, respectively: ^{99}Tc, ^{147}Th, ^{237}Np, natural uranium, enriched uranium, depleted uranium, ^{239}Pu, or ^{241}Am. The phantom is described in detail in section 4.7.1 on page 55.

- The Fission Product Phantom is an anthropomorphic torso phantom similar to the LLNL Torso Phantom with a pin-hole matrix to simulate arbitrary nuclide depositions in the thyroid, lung, lymph nodes, stomach, pancreas, spleen, liver, and kidneys. It includes the head and neck, the complete torso and stub legs articulated at the hips, therefore it is suitable for a sitting position. Source pins are available for ^{60}Co, ^{90}Sr, ^{137}Cs, ^{238}U, ^{239}Pu, and ^{241}Am.

- Two anthropomorphic head phantoms with a homogeneous ^{210}Pb deposition, one on the inner and one on the outer surface of the cranial bone.

- Anthropomorphic head phantom with a homogeneous ^{241}Am deposition on the inner surface of the cranial bone.

Phantoms used for detector efficiency calibration of the WBC are the following:

- The bottle phantom according to Prof. Dr. Schmier (BfS[2]) consists of cylindrical 1-L and 2-L polyethylene bottles filled with a radioactive solution (see also figure 5.3 on page 89). Radionuclides used in this bottle phantom are ^{40}K, ^{137}Cs, and ^{60}Co. At IVM there is a set of bottles with ^{40}K. Ten different configurations are defined representing persons of different weights and heights.

- IGOR is a calibration phantom for the WBC. It is made of polyethylene cuboids and was developed at the Scientific Research Institute for Industrial and Sea Medicine in St. Petersburg, Russia (Research and Technical Centre "Protection", 1997; Kovtun *et al.*, 2000). Each cuboid has holes and can be spiked with small plastic tubes filled with the radionuclide, so called *source rods*. The user needs to build the phantom on the examination table or chair. Six different configurations are defined. The configurations represent children, adolescents, and adults with body weights from 10.6 kg to 95.2 kg in supine, standing, and sitting position. See also figure 5.1 on page 85.

3.3 Efficiency Calibration Methods

In the following sections a selection of calibration methods are shortly described. They are written down in details in standard operating procedures (SOP) (Cordes, 2007) and technical documents (Research and Technical Centre "Protection", 1997). Furthermore, the determination of the chest wall thickness, which is relevant for liver and lung counting, is introduced and the evaluation of the activity is described.

---

[2]Bundesamt für Strahlenschutz (German Radiation Protection Agency)

3.3.1   Calibration of the WBC with the IGOR Phantom

The IGOR phantom is borrowed occasionally from the BfS. The BfS performs intercomparisons with in vivo laboratories to check the quality of the detector efficiency calibration of WBCs.

The source rods are available with many different nuclides from the BfS. The number of rods determines the total activity. The number and position of the used cuboids depend on the configuration. The efficiency calibration needs to have a preceding valid energy calibration on the same day and a background measurement in the same week. The detector response in counts can be determined as the integral of counts in the ROI minus the normalized background counts. Then, the detector efficiency for each radionuclide can be determined with equation 3.2.

3.3.2   Calibration of the PBC with the LLNL Torso Phantom

The LLNL Torso Phantom is used to calibrate the PBC. It can be loaded with radioactive lung lobes and livers, hence it can be used for calibrating detector efficiency for lung and liver measurements. An SOP (Cordes, 2007) describes the standard positions of the phantom and the detectors for the calibration and routine measurements. Concentric markings on the phantom help to find the liver or lung detector positions, respectively. The desired radioactive organ with the desired nuclide is inserted. Again, as for the IGOR, a valid energy calibration and a background measurement is needed. The detector response in counts can be determined as the integral of counts in the ROI minus the normalized background counts. The whole procedure is repeated with a set of four different chest overlays to simulate in total five chest wall thicknesses (CWT):

- no overlay for 12.8 mm,

- overlay 1 for 18.4 mm,

- overlay 2 for 24.5 mm,

- overlay 3 for 28.8 mm, and

- overlay 4 for 35.8 mm CWT.

Three sets of overlays are available. Each set represents different adipose/muscle tissue ratios (0/100, 50/50, and 87/13). In the end of the calibration, plots are generated and linear regression curves ($\eta_{d,l,u}(CWT) = m \cdot CWT + b$) are determined that represent the dependency of detector efficiency η on CWT for each set of overlays, radionuclide and organ (see also figure 7.5.

3.3.3 Determination of CWT

It is well known that CWT is an essential anthropometric parameter for lung and liver counting, but it is not as easy to determine this parameter as for example body weight or height. At IVM, CWT is estimated for each test person with the empirical equation 3.4 (Mohr and Breustedt, 2007). The subject's mass (body weight), height, and chest circumference are required for this formula.

$$CWT = 4.57 \, \frac{cm^2}{kg} \cdot \frac{M}{H} + 0.44cm \tag{3.4}$$

with

$$CWT = \text{chest wall thickness in (cm)}$$
$$M = \text{mass of test person in (kg)}$$
$$H = \text{height of test person in (cm)}$$

The formula was introduced by Doerfel *et al.* (2006). It is based on on investigation performed by Fry and Sumerling (1980) and Dean (1973). They used ultrasound measurements to determine CWT and correlated the values with the ration of body weight and height.

3.3.4 Evaluation of the Activity

The detector efficiency is taken from the corresponding calculated linear regression curve. Then, the activity can be calculated with the number of counts in the ROI by using equation 3.2.

For partial body counting with the phoswich system, the counts from the net peak area, not from the total peak area, in the ROI are taken for the calculation of the activity, since scattering and the body proportions play an important role for low-energy photon emitters. The method is described in detail by Mohr and Breustedt (2007) and was also published by IAEA (1996). In summary, reference spectra of test persons are chosen from a database. They are averaged, normalized, and then subtracted from the measured spectra to obtain the net peak area. The reference spectra are recorded from test persons who are known to be without incorporations, besides natural occurring radionuclides. Such spectra are chosen that are from people with similar anthropometric data (weight, height, chest circumference). An example for this sample procedure is given in the method description by Mohr and Breustedt (2007).

3.4 Problems with Phantom-Based Calibration

The uncertainty in detector efficiency calibration is known to be quite high in part because that only a limited number of physical phantoms are available (Palmer *et al.*, 1989; Kramer and Capello, 2005). Ideally, every individual needs its own phantom. For example, a major concern is the large variety of possible breast sizes and shapes of women. Currently there are no physical female phantoms available, due to the fact that a female phantom would be difficult to standardize. Section 6.2.8 gives more details on this.

Equation 3.4 was introduced by Doerfel (Doerfel *et al.*, 2006). The correlations of empirical anthropometric formulae are generally poor. A test person may be have an atypical CWT for his body mass and height. An alternative for using such formulas are ultrasound measurements. This method is applied at the Human Monitoring Laboratory (HML) in Canada (Kramer, 2008). The accuracy of ultrasound measurements depends on the person, who performs the measurement. The ultrasound probe must be exactly placed to a defined position and some anatomical knowledge is necessary to generate reproducible results. Also the detector position must be adapted for calibration and measurements. A comparison of effort and benefit of ultrasound measurements to determine the CWT lead to the conclusion that the IVM continues using the approach with the empirical formula.

Phantoms do not breath. The effect of breathing on detector efficiency is not fully investigated. There have been experiments with four-dimensional virtual phantoms for radiation treatment planning (Zhang *et al.*, 2006). This virtual breathing is based on an appointed rhythm, which may differ to a real situation. It is still not investigated how a test person breathes in terms of deepness, frequency, and regularity during the measurement. At the moment, one assumes the breathing motion can be approximated with a fixed average position of the chest and the lung lobes.

The accuracy of measurement results depend on the positioning of the detectors relative to the test person. The operator in the PBC needs to reproduce the same detector position for the calibration measurements with the phantom and of course for each test person measurement. Section 6.1.1 goes into details about deviations resulting from positioning.

Chapter 4

Numerical Efficiency Calibration

In this chapter a numerical alternative for classical efficiency calibration is introduced, first, by presenting some basics about Monte Carlo codes, then by describing the new developed simulation environment including used models and software tools. Additionally, the creation of a new voxel model is described.

4.1 Monte Carlo Codes

As already mentioned in section 1.2 radiation transport codes offer a comprehensive range of application for radiation transport problems. They can be used for simulating the WBC or the PBC and hence to determine detector efficiencies. Once the simulation is validated it can be used with a virtual human model to produce calibration data that can be applied for real measurements.

A couple of codes based on the Monte Carlo method are available, partly commercial, partly as open source distributions. A few important examples for such codes are the following:

EGS/EGSnrc is a software package for simulation of coupled electron-photon transport.The current energy range of applicability is considered to be 1 keV to 10 GeV. Originally it was developed under the name Electron Gamma Shower (EGS) at the Stanford Linear Accelerator Center (SLAC) for high energy physics applications. It has been extended for lower energy applications (Kawrakow, 2000; Nelson *et al.*, 1985).

PENELOPE (<u>Pe</u>netration and <u>E</u>nergy <u>Lo</u>ss of <u>P</u>ositrons and <u>E</u>lectrons) performs simulation of coupled electron-photon transport in arbitrary materials for a wide energy range, from a few hundred eV to about 1 GeV (Salvat *et al.*, 2006). It was developed at the University of Barcelona and has advanced methods for electron transport.

Geant4 (<u>Ge</u>ometry <u>an</u>d <u>T</u>racking) is a toolkit for the simulation of the passage of particles through matter. Areas of application include high energy, nuclear and accelerator physics, as well as studies in medical and space science (Agostinelli *et al.*, 2003; Allison *et al.*, 2006) and was originally developed at CERN[1]. This open source toolkit offers flexible usage and educational advantages.

MCNP (Monte Carlo N-Particle) was solely used in this work and is described in detail in following section 4.2.

In addition, there are also deterministic codes, like the commercial **Attila** (Loughlin *et al.*, 2005). This was developed to provide a design based environment for radiation transport simulations and supports computer aided engineering (CAE).

4.2 MCNP/MCNPX

The ISF focuses on the most likely widespread Monte Carlo radiation transport code MCNP. It was developed at Los Alamos National Laboratory in New Mexico and its roots go back to the 1940s. In those days the scientists von Neumann, Ulam, and Metropolis developed Monte Carlo methods that not only made possible the design of nuclear weapons, but also the solution of many other scientific problems

---

[1]Conseil Européen pour la Recherche Nucléaire

(LANL, 2009). MCNP can be used for neutron, photon, electron, or coupled neutron/photon/electron transport. It should be mentioned that only electrons and photons are considered in this work. Areas of application include, amongst others radiation protection and dosimetry, radiation shielding, medical physics, nuclear criticality safety, detector design, accelerator target design, and fission and fusion reactor design.

Basically two versions of MCNP exist at the moment: MCNP5 and MCNPX. For all simulations in this investigation MCNPX (Monte Carlo N-Particle eXtended) in version 2.5.0 (Pelowitz, 2005) was used, otherwise it is mentioned explicitly. Some differences about MCNPX 2.5.0 and MCNP5 1.40 should be mentioned. MCNPX is more suitable for the problems investigated in this work, because its algorithms are optimized to deal with voxel models, for instance the initialization time is significantly shorter. Furthermore, some features of MCNPX, like the anti-coincidence option, are not available in MCNP5, yet. Apart from some details, the two codes are similar suitable for solving the radiation transport problems investigated in this work. According to the developers of MCNP5 and MCNPX, the codes are planned to merge into one single code in the near future. MCNPX with the particles used and within the energy range used in this work has been well tested by various scientists. Therefore it is assumed that MCNPX is able to to produce valid results, as long as the user operates the code correctly (Hughes, 2007).

MCNPX is a particle transport code and its principle is shown in figure 4.1. The user defines the number of particles. The energy of the particles, if not mono-energetic, are chosen by a pseudo-random number generator according to a list of allowed energies and corresponding probabilities (emission spectrum) given by the user (e.g. table 2.1). If the number of particles is large enough, the energy spectrum of the starting particles converges to the given spectrum as entered by the user according to the law of large numbers. This is the concept of the Monte Carlo method. The same effect is used for random sampling a direction of the particle. An isotropic distribution is assumed by MCNP by default, but can be altered by the user if needed. If enough particles are simulated, then the distribution of the simulated particle directions is converging to isotropy. The particles travel through a user-defined space and interact with matter. The user is asked to limit the space and define a volume of interest. As soon as a particle leaves this volume, its interactions are not longer tracked. Besides volume thresholds, there are energy limits as well. As soon as particle energy drops, due to interactions below a certain limit, the particle tracking is stopped. If this is happening inside of a detector volume that tallies energy depositions, the remaining energy of this particle is considered.

Cross-section tables represent the probability functions for particle interactions. Again, the law of large numbers states that the behavior of many thousand particles converge to the real behavior observed in nature. Specific cross-sections for the interactions are calculated from parameters as the composition of the isotopes and the density of matter. The raw data of the cross-sections are derived from the ENDF[2] database, which is implemented in the MCNPX code. The cross-sections are available for each kind of possible interactions. Appendix G "Cross-section libraries" of the MCNPX manual (Pelowitz, 2005) gives details about the implementation of the ENDF data in MCNPX. Basic information that is important for understanding is given in the following paragraph.

The important photon interactions for this investigation that are simulated by MCNPX are Thomson scattering, Compton scattering, photoelectric effect, and pair production for photons with an energy larger 1022 keV. In this investigation electrons play a small but not negligible role. In problems discussed in this work, electrons and positrons are produced as secondary particles after photon interactions. Electrons are far more complex to simulate, since they emerge in large numbers. The electron transport algorithms and data in MCNPX were adapted from the ITS code (Halbleib *et al.*, 1992). The ITS code does not simulate single electron interactions. These are summarized and averaged by the so called *condensed history algorithm*. This algorithm goes beyond the scope of this work and will not be discussed. Summing up, contributions from electrons to resulting spectra are mainly expected in the low-energy range by Bremsstrahlung effects and in case of pair production an annihilation peak will occur.

The hands-on concept of the code is to give an input file as an argument when calling the executable of MCNPX. The input file is

---

[2]Evaluated Nuclear Data File by the National Nuclear Data Center

a kind of script, which tells MCNPX everything about the geometric cells and their mathematical representation, materials, densities, other properties of matter, source particles, physical properties and modes, and tallies (= counters), detectors that output the results. The syntax of this input file is described in the manual (Pelowitz, 2005) and more application-oriented in the theses of two former ISF-students Sessler (2007) and Sohlin (2005). This study will give not give in-depth details on the syntax of the input files. A shortly commented example input file can be found in the appendix C.3 (page 213) of this work.

The time a simulation needs depends on the mainly on the hardware, the physical mode chosen, the number of particles (and hence the statistical quality), their energy and the overall geometric complexity. In this work, a typical simulation of the phoswich detectors in the PBC with 10 million low-energy photons emitted from ^{241}Am) in the lung of a voxel model of 3.4 million voxel took roughly about two hours on a PC[3], while the same setup as an high energy problem with radionuclides like ^{60}Co took about one full day. After the simulation is finished an output file is generated that contains by default all relevant information including the input file itself and the cross-section libraries used on the problem. Also it contains the results of the tallies and statistical tests that MCNP performs automatically to point out the quality of the results in terms of statistics.

---

[3]DualCore Intel(R) Pentium(R) 4 CPU at 3.4 GHz with 3 GB of RAM, running Microsoft Windows XP SP2

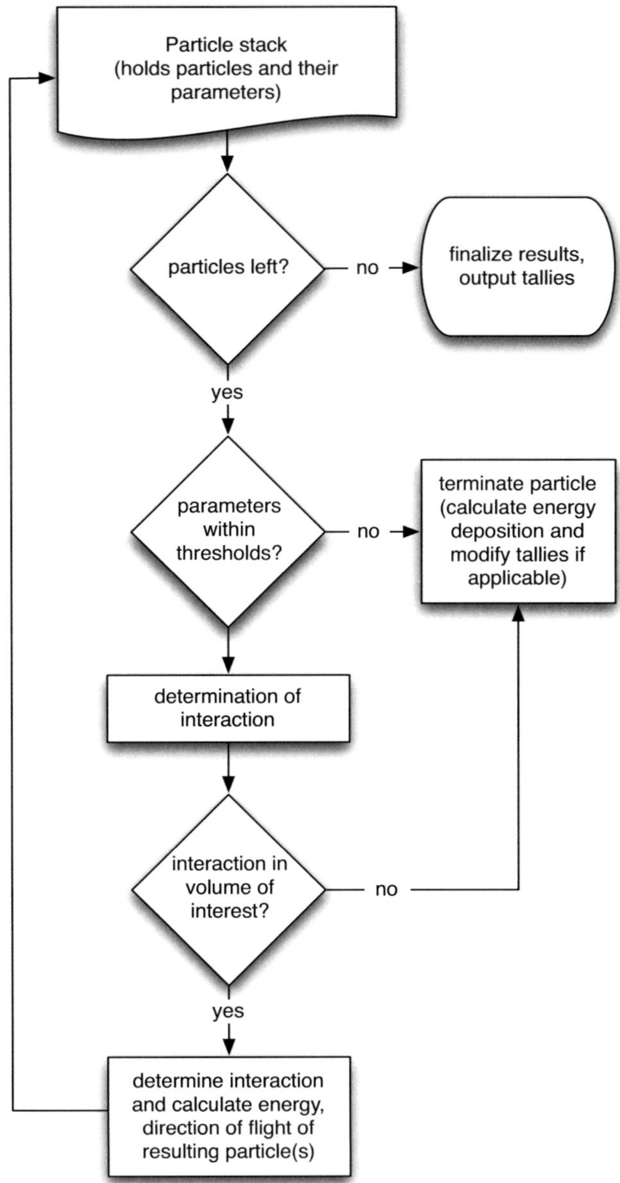

Figure 4.1: Principle of particle transport in MCNPX.

4.3 Efficiency Calculation with MCNP

All simulations presented in this work were performed using MCNPX with default physics features in the `mode e p`, which means photons and electrons are simulated. The cross section libraries *mcplib04* for and *el03* have been used exclusively.

Scintillation and semiconductor detectors as used in the IVM are defined with MCNPX's `F8` tally. The `F8` tally provides the energy distribution of pulses created by radiation in a detector. Energy is accumulated from all tracks of a particle's history. This tally is different to other tallies available in MCNPX, because rather than tallying the particle energy at the time of scoring, the numbers of pulses depositing energy within the bins are tallied. In the appendix C.3 on page 213 there is an example input file with phoswich detectors modeled by `F8` tallies. The `F8` card describes the accepted particles and gives the cell number of the detector volume. The bins are defined by the user by the `E8` card. Bins are equivalent to the channels of an MCA[4].

Representing small numbers in a computer is difficult. Single events may be negligible small, so their energy depositions are rounded to zero. A large number of such events might cause an imbalance. Therefore, MCNPX uses two specials bins. The *zero* bin catches negative scores. This can happen due to the fact, that knock-on electrons in MCNPX are non-analog, i.e. the energy loss is included in multiple scattering energy loss rates rather than subtracted at single knock-on event, hence knock-on events can cause negative energy pulse height scores. If the *zero* bin becomes relatively large compared to the sum of other bins, the `F8` tally can be considered to be invalid. The *epsilon* bin (from 0 to 10^{-5} MeV) bin will catch scores from particles that travel through the cell without depositing energy. This differentiates zero contributions from particles not entering the cell and particles entering the cell but not depositing any energy. In MCNPX the latter type are loosing an arbitrary energy loss of 10^{-12} MeV and appear in the *epsilon* bin (El Bakkari, 2006).

Another feature is the Gaussian energy broadening (`GEB` card) to simulate the energy resolution of the detector (see section 2.3). The user

---

[4]Multi Channel Analyzer

can add three parameters according to equation 2.2. Anti-coincidence circuits, like for example at the phoswich detectors, can be simulated by MCNPX with the `PHL` card in combination with `F6` tallies. It should be noted that a bug in MCNPX 2.5.0 caused problems with the `PHL` card. The problem was solved in version 2.6.0, which is available at the ISF since January 2009. Therefore, most calculations done for this investigation are made with the old version 2.5.0 without the anti-coincidence feature. The differences resulting from using this feature are small for the investigated source nuclides. This is discussed in details in section 6.2.5.

In the output file the `F8` tally results are listed in columns. The first column is the upper bin energy. The first two bins are the mentioned zero and epsilon bins. The second column represents the counts per number of particle histories (`NPS`) in that bin. The third column gives the relative counting error according equation 3.3. An example of the tally output looks like this:

```
cell   522
      energy
   0.0000E+00    6.98700E-04 0.0120
   1.0000E-05    5.30000E-06 0.1374
   9.7656E-04    4.19800E-04 0.0154
   1.9531E-03    2.40400E-04 0.0204
   2.9297E-03    7.59000E-05 0.0363
   3.9062E-03    3.18000E-05 0.0561
   . . .
```

Detector efficiency $\eta_{m,l,u}$ (equation 4.1) can be calculated from the output of the `F8` tally used for each detector. Similar to equation 3.2, a lower and upper limit determine the ROI. It should be noted that MCNPX normalizes the entered photon emission probabilities. While efficiency is usually calculated in *counts per decay* of an radionuclide, the results from MCNPX are in *counts per* `NPS` and must be corrected by multiplying it with the sum of the emission probabilities as entered in the emission spectrum (e.g. table 2.1) in the `SDEF` card distribution (see also appendix C.3 for an example input file).

$$\eta_{m,l,u} = \sum_{i=l}^{u} \eta_{m,i} \cdot \sum_{j=1}^{n} \gamma_j \tag{4.1}$$

with

$\eta_{m,l,u}$ = calculated efficiency in ROI $[l;u]$ in (counts/decay)

$\eta_{m,i}$ = calculated efficiency of bin i in (counts/photon)

l = lower bin number of ROI

u = upper bin number of ROI

γ_j = emission probability of x-ray/gamma-line j of nuclide in MCNP

n = total number of x-ray- or gamma-lines of nuclide

4.4   Modeling the Counting Facilities at IVM

The simulations need models of all objects available in the WBC and PBC that are relevant for MCNPX's particle transport. These models are generated by boolean combinations of rectangular boxes and cylinders that are filled with respective materials and densities.

4.4.1   Model of the WBC Facility

A coordinate system was defined in the measurement chamber. All models have been placed in this coordinate system relative to the origin. A laser measurement system was used to determine coordinates, plane parameters, and cylinder parameters (vector of axis, diameter, and model point). The data were used to describe the walls, the examination table, and the cylindrical detector housings mathematically. Details of the examination table and the inside of the detectors could be derived from this. Drawings of the four NaI scintillation detectors were assumed to be *as-built* and modeled accordingly with nested cylinders. With all this information it was possible to write an MCNPX input file of the WBC including the chamber and its shielding, the detectors, their housing and holder, and the couch. The tally bins and GEB parameters (see section 4.3) for all four NaI(Tl) detectors have been determined for simulation spectra in order to match real spectra. This work was already described in detail in the diploma thesis of Sessler (2007).

4.4.2 Model of the PBC Facility

The PBC was modeled based on previous work (Doerfel *et al.*, 2006; Sohlin, 2005). A coordinate system was defined and the chamber, the two phoswich detectors on the rack were also measured by a laser tracking system. A box-model nests the chamber in layers of the various shielding materials of the graded-Z shielding. Details of the inside of the detectors are modeled according to technical drawings with nested cylinders. The tally bins and GEB parameters (see section 4.3) by Doerfel *et al.* (2006) for the four detector crystals of the phoswich pair have been improved for simulation spectra in order to reproduce real spectra better. As mentioned above, since January 2009, the anti-coincidence feature of the phoswich detectors is modeled with the new MCNPX version 2.6 (see also section 6.2.5).

The model of the examination table by Doerfel *et al.* (2006) that was a box of water with a density of 0.5 g/cm^3 was improved by a more detailed, realistic model made of a steel frame, plastic material, expanded material and artificial leather. The improvement of the new examination table is discussed later in section 6.2.3.

A crucial difference to the WBC is that the detectors and the examination table are not fixed and can be moved to arbitrary locations in the chamber. The position of the detectors can be recorded due to graduations on all moving parts including angles and distances. The models of the detectors can be placed inside model of the chamber with these parameters. An electronic position recording system was developed for the phoswich detector rack. Section 6.3.1 describes the system in detail. It was developed to improve accuracy and reproducibility, furthermore this electronic system can be connected to the internal network at ISF. Parameters can be read from any PC connected to the network.

4.4.3 Further HPGe-Detector Model

Also one of the new Canberra HPGe-XtRa (see section 2.1) detectors has been modeled, which can be used either for WBC and for PCB. The complete detector was modeled including the head with its HPGe-crystal and the external case. The model is based on the technical drawings from the manufacturer but needed some refinements. This work was performed by first comparing measured count rates with the theoretical count rates, i.e. the results from Monte Carlo simulations with MCNPX. The accomplished adjustments adapted thickness and density of different components of the model (Marzocchi, 2008). Like for all other mentioned detector models, the tally bins and `GEB` parameters of the `F8` tally (see section 4.3) have been determined. Finally a good match of simulation and reality was reached.

4.5 Human Models

The requirement of a human model for radiation transport codes is that it reproduces well the geometry, the attenuation, and scattering properties of a real human being. That is the reason why human models mostly have often been developed to resemble reference humans (ICRP, 1975, 2002). Such models are a compromise and constitute a human being with median anthropometric properties.

There are a number of such human models, which can be used in Monte Carlo codes like MCNPX. A bulky overview can be seen online at the website of the Consortium of Computational Human Phantoms (The Consortium of Computational Human Phantoms, 2009). Human models can roughly be divided into three groups.

4.5.1 Stylized or Mathematical Models

The stylized models are mathematical models. These phantoms are made of simple geometric objects that can be described mathematically by for instance spheres, boxes, ellipsoids, and cylinders. Boolean operations of the single geometric objects define the organs. The advantage of the stylized models is that they consume a low amount of memory.

Simulations with them do not take as much computer time as other models, e.g. voxel models. The models can be modified very easily just by changing the mathematical parameters of the geometric objects. The mass of organs and the total body weight can be adjusted according to reference values according to (ICRP, 1975, 2002). The drawback of stylized models is that they do not resemble realistic human anatomy. Details and anthropomorphic particularities are not reflected. There are tools (van Riper, 2009) available that are able to generate stylized models in the MCNPX syntax.

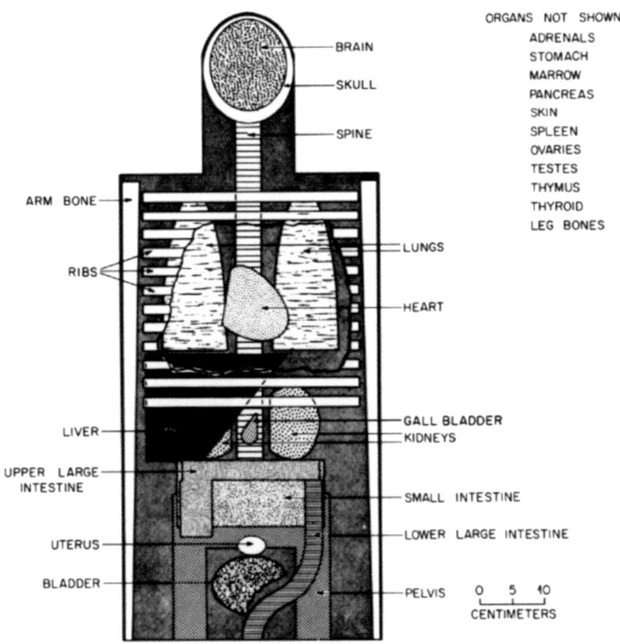

Figure 4.2: Example for a stylized model (Cristy, 1980).

It was Snyder *et al.* (1969) at Oak Ridge National Laboratory (ORNL), who came up with a mathematical model containing organs. They performed external dosimetry studies with these models. Later on, Cristy (1980) developed mathematical models of children. The models created so far were hermaphrodites. That means, that additions of the models defined the gender of the model, while the model base was unchanged. The gender-specific models Adam and Eva (Kramer *et al.*, 1982) have been developed by the GSF[5]. Even now, about forty years later, scientist at ORNL still refining mathematical models, e.g. models with movable arms and legs (Akkurt and Eckerman, 2007).

4.5.2 Voxel Models

Voxel models are models made from CT[6] or MRT[7] scans, and sometimes from photographic data of slices. Just to recall, voxels in a lattice structure are three-dimensional pixels that represent base elements of a voxel model. CT and MRT produce such voxel models. A value that reflects x-ray attenuation or electro-magnetic properties of the scanned tissue is assigned to each voxel. To use such models in Monte Carlo codes, one needs to assign the value of each voxel to specific physical properties. Usually the raw models is segmented into parts that describe organs or tissue of the body with particular densities and materials. This allows to define for instance a source organ or a target tissue for radiation transport problems.

Voxel models are more realistic than the stylized models, depending on their resolution. Small voxel sizes of about one cubic millimeter describe the human anatomy very well. Though, sometimes for radiation protection issues, parts of the human body smaller than 1 mm are important. For instance, for the calculation of the $H_p(0.07)$ skin dose equivalent for external exposures; here a depth of 0.07 mm is relevant. Also trabecular bone structures cannot be resolved adequately from cortical structures at a low resolution. The adsorption of radionuclides in bone tissue depends on such bone structures. Thus, if accurate source distributions in a simulation for internal dosimetry are needed, a high voxel resolution is necessary. The drawback of voxel models is their memory consumption.

Modifying voxel model slices can simply be performed with graphical software. With such software one can simply *paint-over* old voxels with a new *color* (value). The flood-fill feature can reassign a whole

---

[5]Gesellschaft für Strahlenforschung; todays Helmholtz Zentrum München, Germany
[6]Computed Tomography
[7]Magnetic Resonance Tomography

connected area to a new voxel type. This method is tedious, but it can be used to further segment an existing model, e.g. separate left and right kidney. The size of the whole model can be changed linearly by changing the voxel dimensions. Voxels do not have to be cubical. Implementing voxel models in MCNPX was described by Taranenko *et al.* (2005).

It was Caon (2004), who reviewed and discussed twenty-one voxel models. In the following paragraphs several of segmented voxel models that are available at IVM are discussed.

A series of four CT-based phantoms was developed early in the 1990s by **Zubal *et al.*** (1994). First a torso was developed as *The Original Phantom*. It has a voxel dimension of $4\ mm \times 4\ mm \times 4\ mm = (4\ mm)^3$ and consists of $128 \times 128 \times 243 \approx 4.0 \times 10^6$ voxels. Since then, additional body parts have been added. The other three models are called *The Arms Folded Phantom*, *The Arms Down Phantom*, and *The MRI Head Phantom*. They all can be downloaded freely (Zubal, 2009). The torso's voxel size is $(4\ mm)^3$. The authors did not give information about the organ densities and materials. Density and material compositions from the reference male adult (ICRP, 1975, 2002) were taken in case of applicability. For example, one problem was that the sternum and ribs are merged to one voxel type. Since bones and cartilage have different densities and materials, and therefore different attenuation coefficients, it is hard to determine an averaged value for these properties. This limits the appropriate application of the Zubal model series, especially for lung counting applications.

The **MEET Man**[8] (Sachse *et al.*, 2000) was segmented in the 1990s at the IBT of the University of Karlsruhe from image data from the Visible Human Project (Ackerman, 1991). The cadaver was 180 cm high and had a body weight of 92 kg. Ackerman obtained image data of the whole body by CT, while the head and neck was scanned with MRI[9]. Furthermore coronal slices of the rest of the body were also obtained by MRI. In addition the cadaver was photographed in 1-mm-transversal slices. To do this, the body was frozen in a gelatin/water mixture in order to stabilize it. Each slice was ground from the cadaver until it was entirely

---

[8]Models for simulation of Electromagnetic, Elastomechanic, and Thermic Behaviour of Man

[9]Magnetic Resonance Imaging

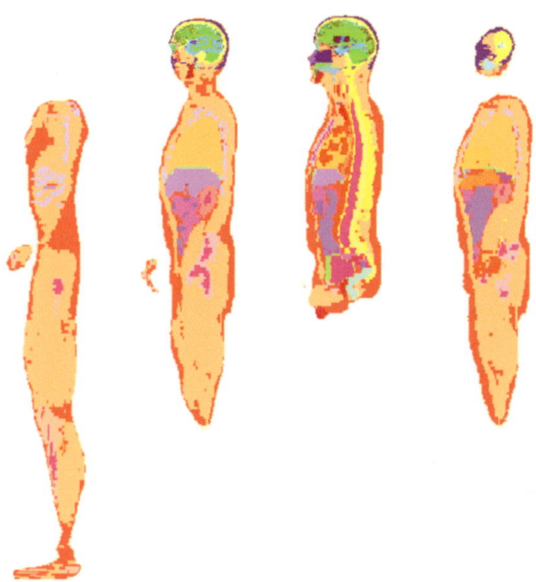

Figure 4.3: Sections of the Zubal model with arms down as a example for a voxel model (Zubal, 2009).

destroyed. The lung lobes of the MEET Man appear to be comparable small. Due to the fact that the image data was collected postmortem, it is assumed that the lung lobes are deflated to a bigger extend than those of a living person.

The ISF has the MEET Man in six different cubical voxel dimensions from $(6\ mm)^3$ to $(1\ mm)^3$ (see table 4.1). Density and material compositions from the reference male adult (ICRP, 1975, 2002) were taken in case of applicability. For example, no difference was made between to two types of bone marrow in the MEET Man models, hence density and composition data of red and yellow bone marrow from the reference male adult needed to be merged and averaged accordingly. The first steps for this work have been performed with the coarsest resolution of this voxel model, due to memory limitations of soft- and hardware.

The data of the VIP-Man were also segmented by Xu *et al.* (2000). They used the voxel model for radiation protection issues years later without knowing the work of Sachse *et al.*. The MEET Man was developed for the purpose to simulate electro-magnetic and mechanical properties in the body.

Table 4.1: Six MEET Man models in different resolutions.

Voxel size	Model dimensions					Number of voxels (10^6)
$(6\,mm)^3$	99	×	56	×	312	1.7
$(5\,mm)^3$	118	×	68	×	374	3.0
$(4\,mm)^3$	148	×	85	×	468	5.9
$(3\,mm)^3$	198	×	113	×	624	14.0
$(2\,mm)^3$	297	×	170	×	936	47.3
$(1\,mm)^3$	594	×	341	×	1873	379.4

Since January 2009 three more voxel models from Helmholtz Zentrum München, Germany (formerly known as GSF) are available at IVM. Godwin and Klara (Zankl *et al.*, 2005) and Frank of the GSF-voxel model family (Petoussi-Henss *et al.*, 2002). Klara and Godwin will be the base of the official ICRP reference voxel model for a male and female model (Zankl *et al.*, 2007; GSF, 2006).

Other important voxel models are NORMAN (Jones, 1996; Dimbylow, 1996), the CNMAN of CIRP[10] (Zhang *et al.*, 2007), and the University of Florida (UF) Pediatric Models (Lee *et al.*, 2006).

4.5.3 BREP Models

A fairly new trend in the world of human models for radiation transport problems are BREP[11] models. Such models use polygon mesh surfaces, NURBS[12] surfaces, or a combination of both to define the boundaries and contours between organs and tissues of the model. The organ meshes originate from commercial databases or segmented tomographic scans. The shape and size of organs can be changed by modifying the surface meshes with CAD software. Knowledge in anatomy is essential for designing BREP models, otherwise the models could be more artistic than scientific.

---

[10]China Institute for Radiation Protection
[11]Boundary Representative
[12]Non-uniform rational B-splines

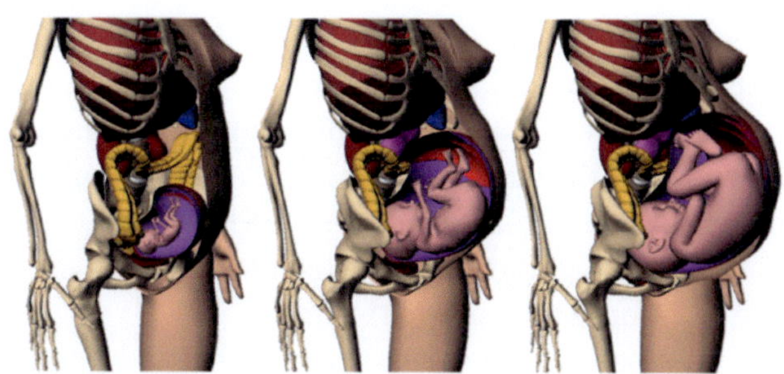

Figure 4.4: Three females in different pregnancy stages by Xu *et al.* (2007) as a example for a BREP models.

Xu *et al.* (2007) introduced a female model in three different stages of pregnancy based on BREP (see figure 4.4).

The same group of scientists at Rensselaer Polytechnic Institute (RPI) are currently developing BREP models called RPI Male- and RPI Female Adult (Xu *et al.*, 2008). The models are deformable and have been assembled from meshes of organs from a commercial anatomical database (Lietzau, 2009). The organ meshes have been deformed to match reference adult values from ICRP (2002) as a starting point for further deformations. In addition, Xu *et al.* (2008) used algorithms for mesh optimization and for avoiding organ overlap (Na *et al.*, 2008).

The RPI Adult models are flexible, which was demonstrated for example by a version of the RPI Female Adult that was modified in breast size (Hegenbart *et al.*, 2008). Brasserie standards have been used to quantify the different breast sizes. The manipulated models with cup sizes ranging from AA to G have been used to calculate detector efficiencies for typical lung counting scenarios in the IVM PBC with two phoswich detectors (see section 6.2.8 for details).

The researchers Zhang *et al.* (2006) of RPI presented a *four-dimensional* model of a breathing torso. In this work, a technique based on NURBS was applied to the surface meshes to generate models of different breathing stages.

Beside the RPI, the UF is an important laboratory to mention in connection with BREP. They introduced the expression *hybrid phantoms* and created adult (Lee *et al.*, 2007a) and pediatric (Lee *et al.*, 2007b) models.

However, current radiation transport codes like MCNPX require voxelized models. To this day, MCNPX cannot handle complicated surface meshes. Voxelization algorithms are used to convert the BREP models into voxel models. For instance, RPI scientists use a method based on parity counting together with the method of ray stabbing on a polygon surface (Nooruddin and Turk, 2003).

Table 4.2 summarizes the general advantages and disadvantages of the three mentioned model types.

Table 4.2: Pros and cons of human model types for radiation transport problems.

Parameter	Stylized model	Voxel model	BREP model
modifiability	good	poor	good
realistic anatomy	poor	good	depends on designer
memory consumption	low	high	high

4.6 Voxel2MCNP

The handling of large data arrays, like voxel models, needs solutions supported by information technology. Important applications, development tools, and toolkits that have been used for this work are listed in the appendix B on page 203. The development of specialized software was necessary to facilitate the handling of voxel models. The software Voxel2MCNP was originally developed to manage voxel models and convert them to the MCNPX-input file format (see appendix C.3, page 213 for an example input file), but now it offers further features and development is ongoing. Brief instructions are given in appendix C.1 that tells the user how to install the software. Also a simple example about how to use the program is described.

The programming language C++ (The C++ Standards Committee, 2009) was used to code Voxel2MCNP. This language is available on most platforms. A high value was set on building cross-platform, open-source tools. This enables future researchers and users at IVM to have the freedom to choose their favorite platform without paying for licenses. Nevertheless, Unix-like platforms have significant advantages for scientific applications.

The code developed in this work was successfully tested on the Unix-based MacOSX 10.4 (Tiger) to 10.5 (Leopard) (Apple, Inc., 2009), Microsoft's Windows XP SP2 to SP3 (Microsoft Corporation, 2009b), and Ubuntu (8.04) Linux (The Ubuntu Community, 2009). The source code was developed with tools mentioned in the appendix B.2 on page 204 and consists of thirty-five source files (`.cpp`) with thirty-four header files (`.h`) and twenty-one Qt user interface files (`.ui`). Table C.1 in the appendix C.2 on page 211 describes the source files and their tasks shortly.

Up to now, almost 20,000 lines of object-oriented code have been written. Voxel2MCNP is designed to be extended in the future and to be used for IVM's routine services. A Doxygen (van Heesch, 2009) documentation was generated automatically. The source code contains explanations, comments, and references at length.

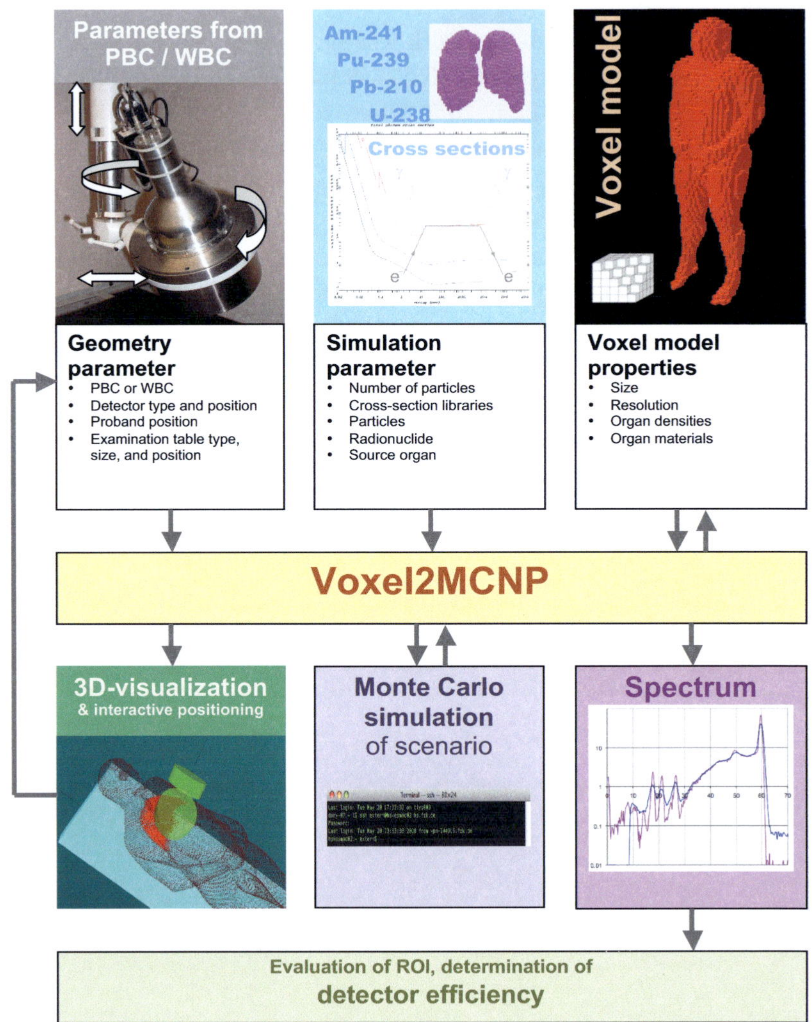

Figure 4.5: Workflow for detector efficiency calculation with Voxel2MCNP.

4.6.1 Parameters

A simplified workflow of Voxel2MCNP is shown in figure 4.5. Geometry parameters, voxel model properties, and simulation parameters are processed by Voxel2MCNP to define an MCNPX input file representing a measurement scenario in the PBC or WBC of IVM. Default parameters and properties are hard-coded, but they can be modified, saved, and reloaded when needed. Geometric parameters are measurement chamber (PBC or WBC), detector type, position of objects (examination table, detectors, voxel model), and their alignment in the chamber. Voxel model properties are the model dimensions, the voxel spacing (size of one voxel), and densities and materials that are assigned to a voxel type. Simulation parameters are:

- MCNPX modes (i.e. particles to simulate),

- choice of the source radionuclide (i.e. particle energy spectrum),

- identification number of the source organ that defines the distribution of the source voxels,

- the number of particle histories to be calculated,

- the cross-section libraries to be used by MCNPX, and

- special features of MCNPX like PTRAC[13] or defining Mesh Tallies (see page 54).

A nuclide library that is based on data from the German national metrology institute (PTB[14]) gathered by Schötzig and Schrader (2000) was created and used throughout this work. Also a material library was built with standard materials like aluminum and steel alloys, detector materials, body tissues, or plastics. The default library can also be modified and saved for future applications. Once all parameters are set, one can let Voxel2MCNP generate an MCNPX-input file.

---

[13]PTRAC particle tracking (Pelowitz, 2005)
[14]German: Physikalisch Technische Bundesanstalt

4.6.2   Viewers

The main window's 2D-viewer (see figure 4.6) shows slices of the currently loaded voxel model on the screen. The slice number can be changed by a slider button or by a spin box. All image slices are created from the voxel model array stored in memory (RAM). Once the model is loaded, one can cycle through all slices quickly without waiting for loading the data from hard drive. The slice plane through the model can be changed by rotating the model in 90° steps.

Voxel2MCNP has an interactive 3D-viewer (see figure 4.7) that is based on the Visualization Toolkit (VTK) technology (Schroeder *et al.*, 2006; Kitware, Inc., 2006, 2009d). The user can navigate through the whole scenario, zoom-in and out, turn on and off objects and organs of the voxel model, as well as changing the opacity of objects. Detector and voxel model parameters for positioning can be changed and checked visually. A virtual red line represents the main axis of the moveable phoswich detector rack. This line helps to position the rack perpendicular to another object or a specific feature of a voxel model (e.g. the sternum). The perpendicular red line is accompanied by a disk that can be moved along the red line. The disk's height relative the origin of the chamber is displayed in the GUI[15]. Points of interest can be located this way in the displayed scenario.

The displayed organs are reduced to the boundary voxels' sides that delimit the organ's volume, which enables a fast 3D-visualization even on slow computers. These organ surface meshes can be exported in the VTK file format and displayed with Paraview. Organs and complete scenarios can also be generated as VRML[16] worlds that can be viewed by browsers with a specific plug-in[17].

---

[15]Graphical User Interface

[16]Virtual Reality Modeling Language(International Organization for Standardization, 2009)

[17]e.g. the Cosmo Player

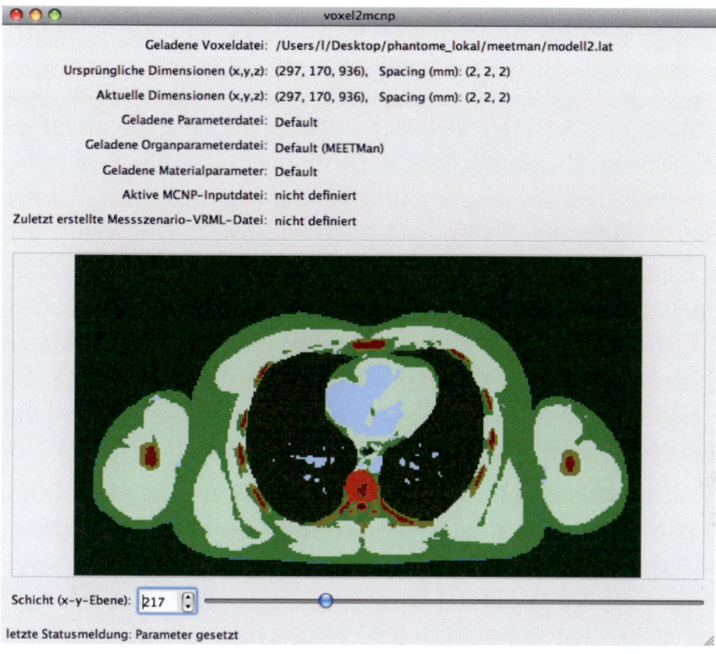

Figure 4.6: Voxel2MCNP's main window with a voxel model loaded.

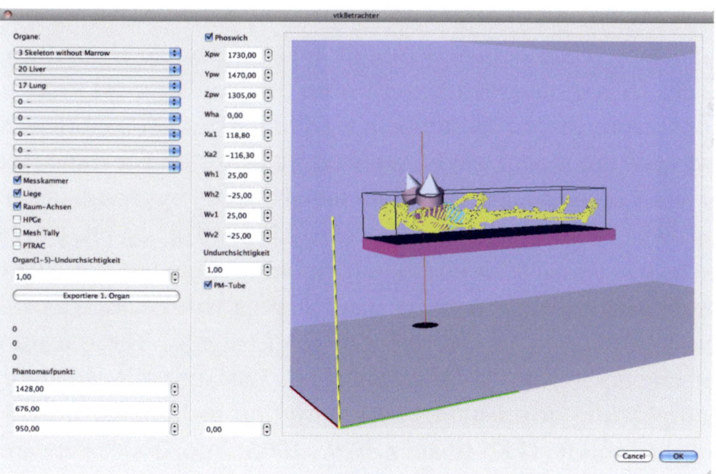

Figure 4.7: Voxel2MCNP's VTK-based interactive 3D-visualization window.

4.6.3 Processing Voxel Models

A focus of Voxel2MCNP is the handling of voxel models. Readable file formats are ASCII-codes (e.g. the Spitz knee phantom used in (Gómez-Ros *et al.*, 2008)), binary formats like simple byte arrays (e.g. Zubal models), binary arrays with a header (e.g. kaLattice format (IBT Universität Karlsruhe, 2009)), and BMP[18] bitmap series. While MCNPX cannot handle large voxel models (see also section 4.7.7), Voxel2MCNP can handle at least 380 million voxels: a test on a MacBookPro computer with 4 GB of RAM showed that the finest resolved MEET Man model can be loaded and processed. It is assumed that Voxel2MCNP can handle larger models as long as the computer has enough memory installed and its operating system is able to address the memory.

Voxel models can be rotated in 90° steps around any of the three coordinate axes. The model can be mirrored along all three normal planes. The dimension of a model can be enlarged by adding new voxel layers onto the recent model boundaries, as well as one can cut layers of voxel slices. An automatic cut function cuts away dispensable layers of air. If the model dimensions coincide, separated organs can be added as binary bitmap series to an already loaded phantom. This feature was used in section 4.7.4 to build a phantom from single organ segments.

Organs can be assigned to another organ or isolated from the rest of the model. A region-growing algorithm fills voids (voxel value equals zero) or specific organs by setting a seed to the desired coordinates. The resolution of the model can be reduced by factor two or three (see also section 4.7.7). The models can be stretched linearly by changing the voxel spacing. By exporting the voxel model to a bitmap series, one can modify the model slice-wise with image manipulating programs, e.g. the GIMP. More sophisticated algorithms for manipulating voxel models are dilation, erosion, and voxel-wise movement of single organs. These features are applied and demonstrated in section 6.1.2. Latest developed algorithms for manipulating voxel models are based on technology provided by the Insight Toolkit (ITK) (Ibanez *et al.*, 2005; Yoo, 2004; Kitware, Inc., 2009c). These algorithms are discussed in more details in chapter 7.

---

[18]Windows Bitmap

4.6.4   Voxel Model Input File Syntax

The recommendation of Taranenko *et al.* (2005) have been followed for setting up voxel models in MCNPX. For a detailed input file syntax description, check the appendix C.3. How Voxel2MCNP outputs the code for the voxel model is shown in the following example. Only parts relevant to the voxel model are displayed:

```
c used voxel model <</Users/lars/torso0.lat>>
c voxel dimensionen in [mm] (x,y,z)=(2.8125, 2.8125, 2.4)
999 1 -0.00129 -54 u=999 lat=1 fill=0:169 0:94 0:208 &
```

Comment lines in MCNPX input files start with c. Comments can also be appended after a $ symbol. The cell that holds the phantom is 999, which is also the number of the *universe* (u). The concept of universes is the nesting of cells in each other. Cell 999 is basically made of material 1, which will be later defined as air with a density of 0.00129 g/cm^3. The geometry card number 54 is the original voxel for the model. Later, in this card the voxel dimensions will be declared with the help of a rpp macrobody. The minus sign indicates that the cell is the inside of the macrobody. Cell 999 is a lattice (lat) of type 1, i.e. a hexahedral lattice and is filled with $170 \times 95 \times 209 = 3,375,350$ elements. The ampersand connects the following lines to the cell 999.

```
16 6206r 1 10r 16 154r 1 16r 16 149r 1 20r 16 147r &
16 142r 1 26r 16 141r 1 27r 16 140r 1 28r 16 140r 16 &
16 141r 1 26r 16 144r 1 21r 16 13760r 1 9r 16 155r  &
16 146r 1 23r 16 143r 1 26r 16 141r 1 27r 16 140r 16 &
...
```

These lines represent the voxel array written as universe numbers. For example, 16 stands for the first voxel (universe 16 will be defined later as air) and is repeated for 6206 times. Then 11 times the voxel of universe 1 follows, and so on, until all 3,375,350 voxel are defined, first the x-, then the increasing y-, and finally increasing z-indices.

```
1     50    -1.091    55    u=1
2     51    -1.101    55    u=2
3     52    -1.086    55    u=3
4     53    -1.088    55    u=4
5     55    -1.084    55    u=5
...
```

After the voxel array, cells with the universe definitions of the voxels follow. Here again, the cell number coincide with the universe number, like recommended by Taranenko *et al.* (2005). The second column represents materials, and the third the negative densities. The negative sign for the density tells MCNPX to use the unit g/cm^3. The forth column is a simple plane numbered 55 that defines the half space in which the universes are valid.

```
54   rpp 148.453 148.73425 67.6 67.88125 95 95.24 $ urvoxel
55   pz  -300 $ plane for universes
. . .
```

As already mentioned, geometry card 54 is the original voxel. With the parameters of rpp one defines the coordinates of the bounding box in centimeters.The first two numbers define the starting and end coordinate in x-direction, the next two pairs define those for the y- and z-direction, respectively. 55 is the already mentioned plane above which the universes are valid. The plane (pz) is normal to the z-direction at $z = -300$ cm.

With the SDEF card the source is defined. The source cells are declared as distribution D1. The cards si1 and sp1 are associated with D1. They define a uniform distribution of all source voxels. One source voxels is delimited with brackets. The greater-than signs describe the universe order and the voxel indices in the lattice are in the squared brackets. Again, like in the array definition above, the ampersand connects all lines. In this example 195,757 voxel had been defined as source voxels.

```
SDEF SUR=0 CEL=D1 ERG=D2 PAR=2 X=D3 Y=D4 Z=D5 EFF=1e-4
si1 L (19<999[73 85 107]<531) (19<999[69 85 108]<531) &
(19<999[70 85 108]<531) (19<999[71 85 108]<531) &
(19<999[73 85 108]<531) (19<999[74 85 108]<531) &
. . .
(19<999[60 9 123]<531) (19<999[61 9 123]<531)
sp1 1 195756r
. . .
```

The distributions D3, D4, and D5 are defined column by column that is indicated by the hash sign. While the si cards define the coordinates already known from the original voxel, the sp cards with the entries 0 1 constitute a uniform distribution. Together this part describes a volumetric uniform distribution inside the original voxel.

```
#   si3         sp3  si4         sp4  si5       sp5
```

```
    148.453    0    67.6       0    95      0
    148.73425  1    67.88125   1    95.24   1
...
```

Note: A memory saving alternative of defining all voxels separately in the `sil` list is using

```
si1  L  (19<999<531)
sp1  1
```

after the `SDEF` card to define all organ (`u=19`) voxels at once. The drawback of this alternative is the speed of the simulation, that decreases a few percent and it is less flexible. Therefore, the first method mentioned was used for Voxel2MCNP.

4.6.5 Other Features

For his diploma thesis Sessler (2007) contributed a C++-class for implementation of a model of the IGOR phantom (mentioned in section 3.3.1). The model of IGOR can be shifted on the examination table in any direction. Besides voxel and IGOR models, point source input files can be generated.

Another feature of Voxel2MCNP is the analysis of voxel models. Basic model information are displayed on the main window (figure 4.6), while more detailed information can be generated by choosing the *Organstatistik* button in the *Analyse* menu. This will give detailed information for each organ, i.e. voxel number, volume, assigned density, organ mass, and the organ's center of gravity. The data can be displayed on the screen or saved in the Excel readable CSV[19]-format.

For the evaluation of the data contained in the MCNPX-output files, Voxel2MCNP can extract the spectra from the `F8`-tallies and save them in the CSV-format for further processing in plotting tools. Excel file templates have been used to quickly evaluate the data, display spectra, calculate the detector efficiency in a specified ROI.

A special mode of MCNPX generates the already mentioned `PTRAC`-files. These log particle information at run-time in ASCII or binary format. Voxel2MCNP can read these files and generate a three-dimensional

---

[19]Comma-Separated Values, an ASCII file format

visualization (see figure 4.8) and save it VRML format. This feature is useful to check whether the source definition is correct.

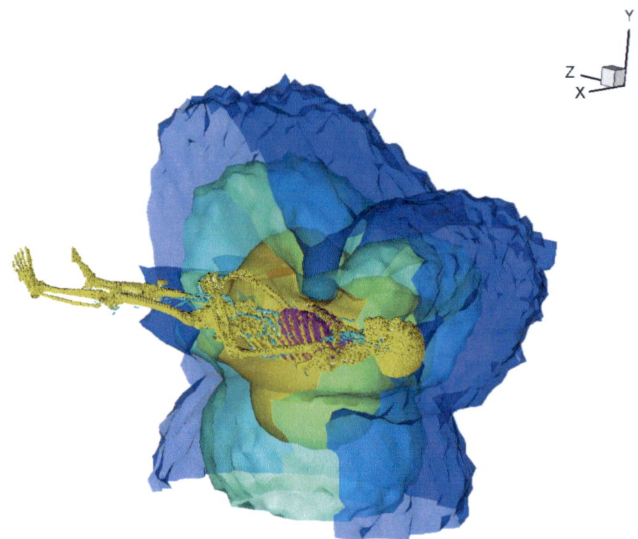

Figure 4.8: 3D-representation of iso-flux surfaces (clouds) around the MEET Man's radioactive lung. The skeleton and the lung lobes are superimposed to the surfaces for better orientation. Funnel-like features can be seen over the lungs and the head. They are caused by the attenuation of the phoswich detectors placed on the chest and the skull of the model. The coarseness of the outer surfaces are caused by statistical noise.

Another useful feature is the conversion of MCNPX's Mesh Tallies into VTK-compatible format. With Mesh Tallies, for example, the photon flux in a volume can be quantified in single cubical elements. MCNPX offers just a 2D display of the Mesh Tally results. Iso-flux surfaces can be displayed in 3D by Paraview. Mesh Tallies can be used, for instance, to optimize detector positions.

4.7   Creation of a Voxel Model

To be able to compare measurements involving a calibration phantom with Monte Carlo simulations, a model of the measurement scenario including a model of the phantom of the is necessary. Therefore, a voxel model of the LLNL Torso Phantom[20] was created. The phantom belongs to IVM's calibration equipment.

The first step of the creation of the voxel model of the LLNL Torso Phantom was to scan the phantom by computer tomography and to generate a three-dimensional image – an unsegmented voxel model – in the common DICOM format (NEMA, 2008). The open source software OsiriX (Rosset *et al.*, 2004) was used to view, evaluate, and segment the data. Segmented organs then have been exported as binary bitmaps and later imported in Voxel2MCNP (section 4.6) to put all organs together to generate a complete voxel model and saved in series of BMP-files[21]. Manual, slice-wise post-editing of possibly twofold defined voxels was performed with the GIMP (see also appendix B.1, page 203).

Besides the segmentation, one needs to specify the material and the density of each voxel for the Monte Carlo code in order to determine the cross section of the voxel matter. For the LLNL Torso Phantom, some values could be taken from literature or measurements. Since the manufacturer's documentation was inadequately, most values needed to be determined or redetermined for verification purposes.

Large voxel models have a high demand of computational power. First, they consume a lot of memory and secondly, as an effect of this, they need large processor resources for each mathematical operation. Therefore, a coarsening algorithm was developed, tested, and later, applied to the voxel model to handle the large amounts of data in MCNPX.

The following subsections explain the procedure of the mentioned steps of the creation of a voxel model in detail.

4.7.1 LLNL Torso Phantom

Amongst other scientists Anderson *et al.* (1979) identified several problems associated with the calibration of in vivo measurement systems for the assessment of actinides, which are described in details by Taylor (1997). This was a motivation for Griffith *et al.* (1986) to develop a prototype of a phantom at the Lawrence Livermore National Laboratory

---

[20]Lawrence Livermore Realistic Torso Phantom (see also section 3.2)

[21]Microsoft Windows Bitmap format, version 3

(LLNL). The phantom prototype represents a male adult with a height of 177 cm and a body mass of 76 kg. The IVM owns a copy of the LLNL Torso Phantom since 1985. It was manufactured according to the specifications of the LLNL prototype and commercially distributed by Humanoid Systems, Incorporated.

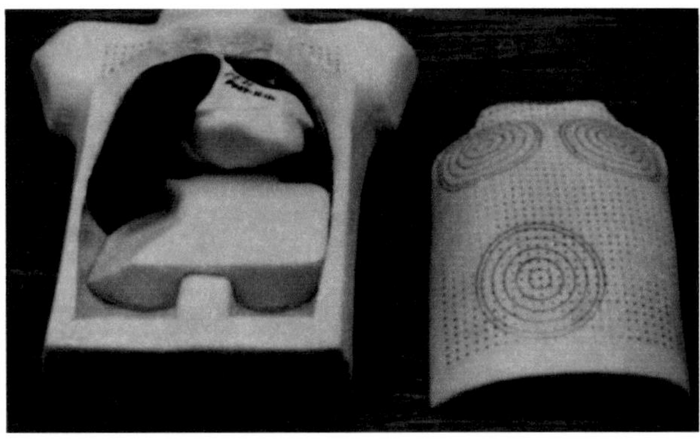

Figure 4.9: Partly disassembled LLNL Torso Phantom (Hickman, 2005).

The LLNL Torso Phantom consists of organ parts made out of tissue-equivalent synthetic material. The materials are solid or foamed polyurethane mixtures. The bones of the first prototypes were real bones from human cadavers. Later, when the phantom became a commercially available product, the bone tissue was substituted by polyurethane mixtures (Griffith, 1980). The content of $CaCO_3$ (calcium carbonate) in the polyurethane mixture was varied to produce similar attenuation coefficients like the target organs. For instance, this was used to produce substitute tissue with a certain ratio of muscle versus adipose tissue (see Ca content of materials in table 4.3). Later, Hickman (2005) wrote an illustrated manual how to assemble and disassemble the phantom parts (figure 4.9). The parts (see also table 4.4) are the following:

Torso is the part of the phantom containing all other parts. In the torso, there are back parts of the cast-in ribs and vertebrae.

Torso cover covers the inner organs of the phantoms and contains the front part of the cast-in ribs and the sternum.

Left and right lung lobe are the most superior organ parts and are made out of foam with a low density.

Lymph node block sits between the lung lobes and supports the heart.

Heart sits between the lung lobes.

Upper kidney block is positioned posterior under the heart and liver.

Lower kidney block is positioned inferior to the upper block and fills the lower back-part of the torso.

Liver is anterior to the upper kidney block and is supported by the so called liver envelope.

Liver envelope is anterior to the kidney blocks and incloses the liver.

Sets of overlays of four different thicknesses (representing chest wall thicknesses from 12.8 mm to 35.8 mm) and three different materials can be applied separately onto the torso cover.

The manufacturer provided information about elemental composition (table 4.3) of the polyurethane mixtures used. The majority of the parts are made of material C - muscle tissue. The casted-in bones are composed of material E, while the lung lobes are made of the foamed material F. The overlays are available in the materials A, B, or C (muscle/adipose substitutes). According to the manufacturer's description, the D material (cartilage) is used to cover the inside of the torso cover to represent the sternum. Unfortunately, no specifications or detailed information about masses, densities, and volumes of the organs are available. Some values could be calculated with data from sample reports about the material charges that have been shipped with the phantom. The material charges are connected to the serial number of the organ. Unfortunately, there are

Table 4.3: Types of tissue-equivalent polyurethane mixtures used in the LLNL torso phantom according to the Technical Data and Instruction Manual (Radiology Support Devices, 1985).

Code letter	Tissue substitute	Elemental composition (mass %)				
		H	C	N	O	Ca
A	13% muscle / 87% adipose	9.44	61.60	3.57	25.40	0.00
B	50% muscle / 50% adipose	9.24	60.80	3.85	25.40	0.78
C	Muscle	9.03	59.40	3.30	26.60	1.70
D	Cartilage	8.89	61.10	2.98	24.70	2.33
E	Bones	6.38	47.20	2.12	31.30	13.00
F	Lung	8.00	60.80	4.20	24.90	2.10

no sample reports on all material charges used to build the the phantom. Therefore it was decided to weigh all parts (tables 4.4 and 4.5) and to calculate the volume of each part after the segmentation by number of voxels times the voxel volume.

Supplementary, there are radioactive organs (table 4.5). The IVM owns ten pairs of lung lobes and four livers containing radionuclides. The manufacturer assures a uniform distribution of the radionuclides in these organs.

4.7.2 Computed Tomography

In November 2006 the LLNL Torso Phantom was scanned with a Siemens *Somatom Sensation 16* computer tomography system at the Vincentius Kliniken in Karlsruhe.

The device automatically generates image slices from the helical scan movement. The phantom was scanned completely and 633 images slices with a thickness of 0.8 mm have been produced. Each slice had 512×512 voxels with a resolution of 0.9375 mm $\times$ 0.9375 mm. According to this, one voxel's volume amounts to 0.703125 mm^3. A Hounsfield value was assigned to each voxel resulting in a three-dimensional, 12-bit-grayscale image with $512 \times 512 \times 633$ voxel. The device's attenuation was calibrated in a way that water gives a Hounsfield value of zero. In total, five scans have been made. The phantom torso was kept fixed on the bed and scanned one time without overlay, then one time for each of the four overlays. The scan data was saved in the DICOM format on CD-ROM.

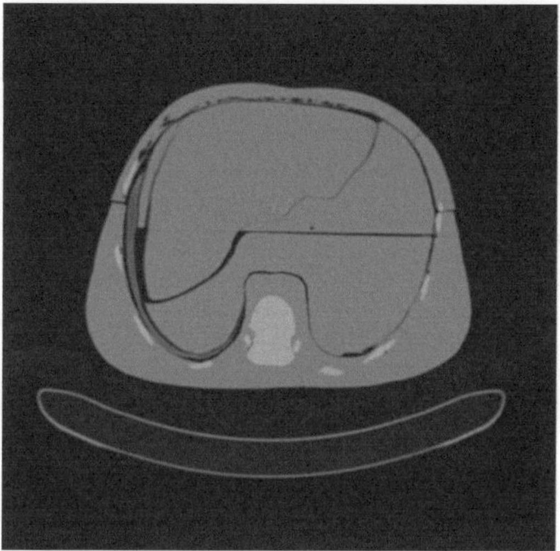

Figure 4.10: Example of a grayscale image of one slice of the CT-scanned phantom. The darker the color appears, the lower the Hounsfield value is. The U-shaped metal frame of the bed is clearly visible below the phantom. The brighter areas in the torso are ribs and a vertebra. Dark air crevices in the phantom simplify the perception of the single phantom parts. For instance one can see the liver sitting in the left-upper sectors of the torso. The medical images are usually viewed from feet to head (inferior-superior), as if the physician stands in front of a patient's bed and imagines the slices in his body. That is the reason why the liver is on the left side in this image. Roundish dark spots represent air bubbles, which can be observed particularly at the inside of the torso cover.

Figure 4.10 also shows the difficulty of displaying $2^{12} = 4096$ monochromatic (grayscale) values at the same time. DICOM viewers let the user control over contrast and brightness, so one can give the structures of interest an appropriate contrast.

4.7.3 Segmentation

Segmentation of the CT-data was performed with the help of OsiriX 2.6 (Rosset *et al.*, 2004). OsiriX enables the user to navigate quickly through the DICOM data and provides various three-dimensional viewing options (figure 4.11).

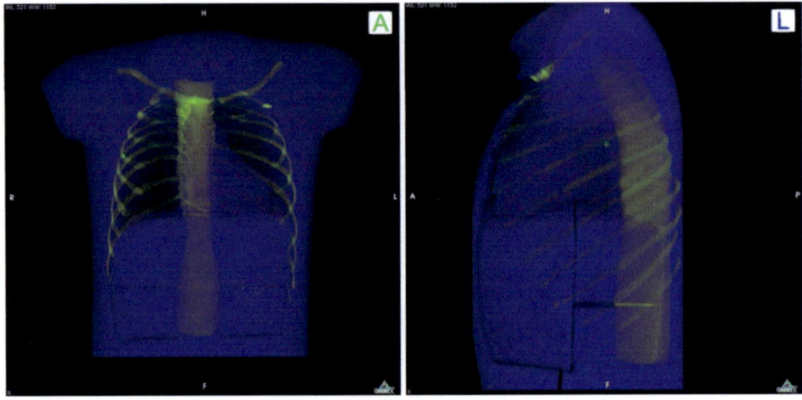

Figure 4.11: Front and side view onto the semi-transparent phantom with one of OsiriX's 3D-viewers. A color interpretation of the Hounsfield values was chosen to enhance the contrast of the different materials.

The segmentation technique used is based on marking structures in the images (figure 4.12) as so called regions of interest (ROI). This *ROI* is referring to a set of voxels and differentiates from the ROI mentioned in terms of energy ranges when analyzing spectra (section 3.1).

The CT-scans of the LLNL Torso Phantom have been segmented into twelve organ segments and the *air* segment. Additionally, four overlays (50% adipose, 50% muscle) have been segmented. The single segments are:

- the torso without bones and vertebrae,

- the torso cover without ribs,

- the left,

- and the right lung lobe,

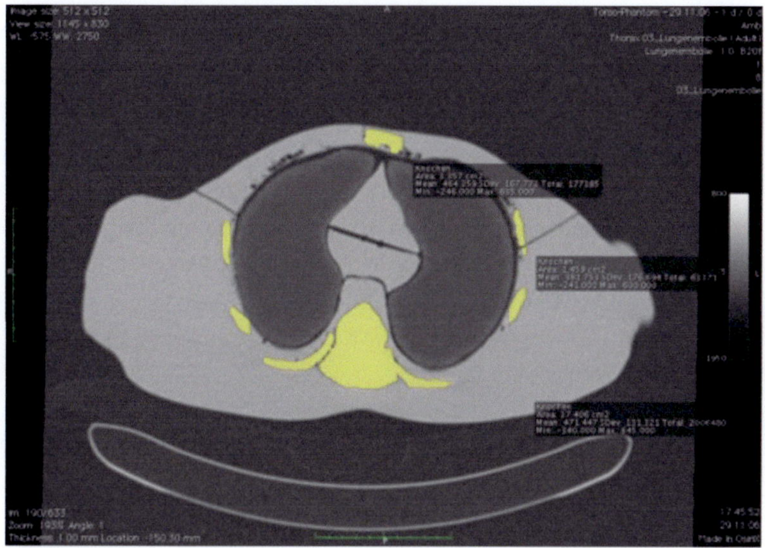

Figure 4.12: Screenshot from OsiriX. Bone substitute material is marked as the ROI in transparent yellow.

- the lymph node block,

- the heart,

- the upper,

- and lower kidney block,

- the liver,

- the liver envelope,

- bones in the torso cover,

- bones in the torso,

- and finally the air that surrounds the phantom, fills the gaps between the parts, or that is trapped in bubbles.

Threshold Segmentation

First a threshold segmentation technique was applied. All voxels with Hounsfield values above or below a chosen threshold can be isolated. The thresholds can be determined by evaluating histograms; the voxels are sorted by their Hounsfield value in small intervals. In histograms, the frequency in each interval is plotted against ascending Hounsfield intervals. OsiriX has a function to display histograms (see figures 4.13 and 4.15). Air and bones are predestinated for this technique, due to their extreme Hounsfield values at the lower and upper end of the histogram scale, respectively.

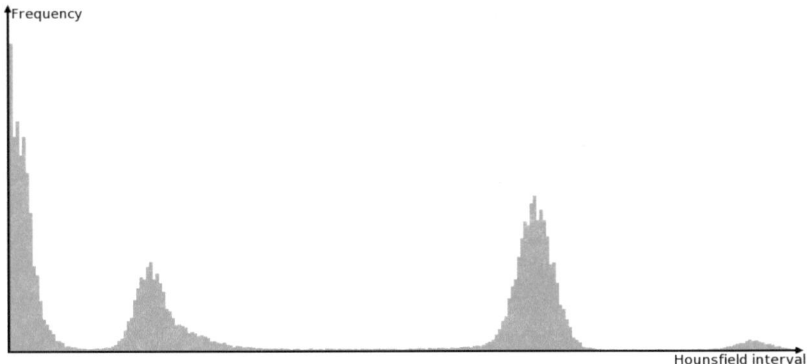

Figure 4.13: Histogram generated by OsiriX. The peak on the very left side represents *air*. The next peak to the right represents lung-, then muscle- and finally on the right end there is the small peak of bone tissue substitute. Evaluating histograms helps to find the thresholds for segmentation.

Air was isolated first. All voxels with Hounsfield values below -800 were marked as ROI and set to the maximum Hounsfield value, while all other voxels have been set to the minimum value. The whole three-dimensional image was now exported as a series of 633 binary (one bit per voxel) bitmaps and converted in the BMP-file format. Such binary bitmap series have been generated from ROIs for all segments mentioned in the list above.

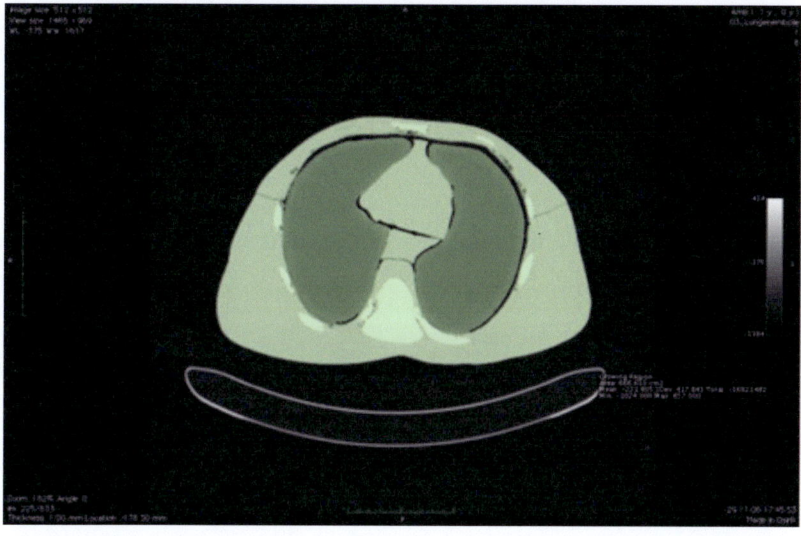

Figure 4.14: A ROI (green coloration) marks all non-air voxel of the phantom slice #225. Figure 4.15 represents the histogram of this ROI.

Table 4.4: Mass of the phantoms parts.

Part / organ	Material code letter and serial number	Mass (g)
Lower kidney block	C2-214	1310
Upper kidney block	C2-214	1911
Heart	C2-214	1173
Liver envelope	C2-214	2688
Liver	C2-214	2211
Lymph node block	C2-214	176
Left lung lobe	F 139	527
Right lung lobe	F 139	687
Overlay 1	A-124-1	1595
Overlay 2	A-124-2	2800
Overlay 3	A-124-3	3860^a
Overlay 4	A-124-4	5200^a
Overlay 1	B-127-1	1380
Overlay 2	B-127-2	2609
Overlay 3	B-127-3	3760^a
Overlay 4	B-127-4	5280^a
Overlay 1	C-131-1	1420
Overlay 2	C-131-2	2650
Overlay 3	C-131-3	3800^a
Overlay 4	C-131-4	5280^a
Torso cover incl. bones	C1-127	3129^a
Torso incl. bones	C1-127	18020^a
Cartilage in torso	D-138	n/a
Bones in torso	E-154	n/a

[a]Weighed with balance Soehnle S10-2720 (max. 60 kg; min. 0.4 kg; e=20 g for m<30 kg; e=50 g for m≥ 30 kg). All other organs weighed with Sartorius 3713MP (max. 3000 g; e=0.1 g).

Table 4.5: List of radioactive organs and their masses at IVM for the LLNL Torso Phantom.

Organ	Serial number	Radionuclide	Mass (g)
Liver	533	^{241}Am	2222
Liver	537	^{238}Pu	2234
Liver	539	^{147}Pm	2263
Liver	539	^{239}Pu	2237
	left/right lobe		left/right lobe
Lung	539L / 539R	^{147}Pm	529 / 660
Lung	576L / 576R	^{241}Am	535 / 717
Lung	579L / 579R	^{237}Np	531 / 680
Lung	577L / 577R	^{232}Th	555 / 701
Lung	578L / 578R	^{99}Tc	542 / 681
Lung	580L / 580R	^{238}Pu	479 / 681
Lung	581L / 581R	^{238}U	530 / 714
Lung	582L / 582R	U, natural	529 / 653
Lung	584L / 584R	^{239}Pu	534 / 697
Lung	588L / 588R	^{235}U, 93%	556 / 734

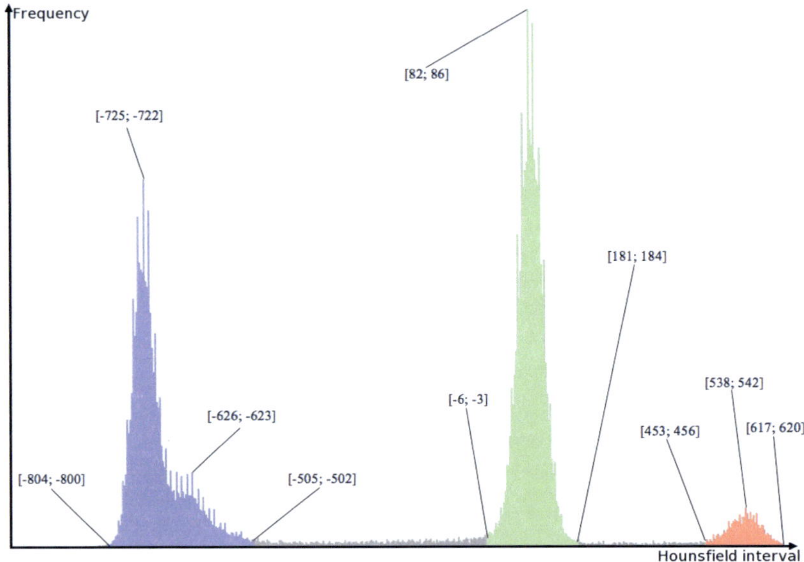

Figure 4.15: A more detailed histogram of the ROI as displayed in figure 4.14. The histogram is based on 75,848 voxels. The intervals of some selected histogram bars are written in squared brackets. The height of a bar represents the abundance of voxels in each interval. The air peak is not displayed, because air was excluded from the ROI. There are three main peaks visible: The first one (blue), representing the foamed lung tissue substitute (F material) at around -724 with a slight tailing to higher Hounsfield values. The next peak (green) is more symmetric and centered at around 84. It represents most of the solid polyurethane (C and D). Finally, the smallest peak (red) at around 540 represents the bone substitute material (E). Cartilage substitute (D)-material seems to be less abundant and could not be separated from the muscle substitute (C). Therefore, the D-material was ignored throughout this report.

Region Growing

For the segmentation of the bones, a second technique has been used: the so called region growing method. The user sets a seed in the image and an algorithm checks neighboring voxels if they are in a predefined Hounsfield value interval. If a suitable voxel was found, it becomes a new seed. This technique works in three dimensions. A suitable range for the bones was from 400 to 700. Once a seed on a rib is chosen, the remainder of that rib gets marked as ROI. The ribs are discontinued at the boundary of the torso to the torso cover, so the region growing stops at the boundary. Therefore, the bones in the torso cover could be separated from the bones in the torso.

The volume of both bone segments was determined by counting the voxels and multiplying them with the volume of a single voxel. A density of 1.268 g/cm^3 for bone substitute material was calculated from the sample reports that have been shipped with the phantom. The mass of each bone segment could then be calculated and subtracted from the weighed values of the torso or its cover, respectively. By doing this, the masses for the remainder tissue of the torso and its cover without bones was determined.

More difficult was the segmentation of the C-material parts. The comfortable three-dimensional region growing technique did not work in this case. The air gaps between the organs were to small that seeds flowed over to the next organ. The problem was solved by using a two-dimensional region growing technique for all relevant slices. In case the air gaps were to small, the ROI was marked by a manual ROI-painting tool. It turned out to be tedious labor, since for some parts the number of slices were numerous. Besides setting thresholds for the region growing, OsiriX's advanced neighborhood threshold mode was used. Not only primary neighbor voxels are checked in this mode, but also voxels in a predefined radius. This mode lowers the tolerance and stops the seeds more easily, which partly simplified the work. The parameters for thresholds and radii have been chosen individually for each organ. Sometimes they even have been modified for single slices.

Since the boundaries of the F-material of the lung lobes are clearly distinguishable, the three-dimensional region growing technique with the mentioned neighborhood mode was used. As an example, the threshold settings for the lung lobes were -900 to -100 with a radius of two voxel, while for C-organs threshold values of -100 to 1000 and radii ranging from two to four voxel have been used. Using neighborhood mode caused the region growing to stop before the organ boundary was reached. To improve the ROI, it was dilated by a few voxels after the region growing without risking flooding the neighboring organ. Here as well, OsiriX provided convenient tools to do this.

4.7.4   Building the Model of the Phantom

Ignoring bone material and marking it together with the torso or the torso cover part was another trick that was used to facilitate the work. Since the bones have been segmented already, they could be added later easily by simply overstriking other segment's voxels when the phantom segments were put together. The same trick was applied for air bubbles. Such logical operations have been performed with binary bitmap images series loaded in Voxel2MCNP. The algorithm of building the phantom model is shown in figure 4.16.

A little subsidence of the phantom into the examination table was observed for the two heavier overlays 3 and 4 (see figure 4.17). Apart from that, the CT system delivered almost perfect reproducible images of all CT-scans with different overlays, so the torso was is a stable position.

The following logical technique was applied to create segments from the four overlays. First the non-air voxels have been segmented with threshold segmentation in both scans, the one with and the one without the respective overlay. The non-air segment of the normal torso has been subtracted from the one with the overlay resulting in the segment of the overlay, which in a further step could be separated from noise and contours caused by the subsidence with the help a simple three-dimensional region growing.

Due to the subsidence of the phantom with overlay 4, its segment needed to be shifted by one voxel in ventral direction with Voxel2MCNP's organ-moving tool before the subtraction was applied. Figure 4.17 illustrates this.

Sometimes a voxel position was defined twice on the boundaries of the segments. In this case a special overlap identification number was introduced (see algorithm in figure 4.16). That is why a final manual step was necessary. All slices have been checked for such overlap voxels with the GIMP software. It was up to the user's visual decision to which neighboring organ the multiple defined voxel was assigned to. Figure 4.18 shows a VRML representation of the final version of the voxel model of the LLNL Torso Phantom.

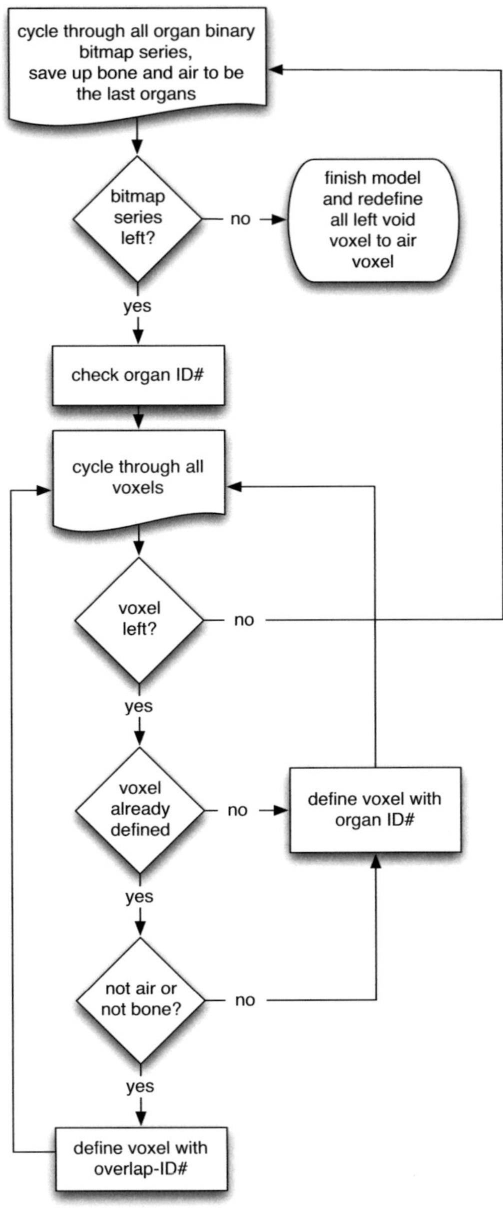

Figure 4.16: Algorithm of building the phantom model.

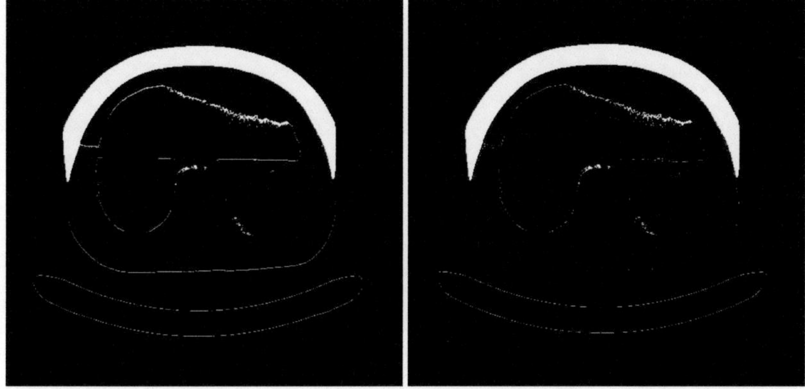

Figure 4.17: Differential images of pre-segmented CT-scans. The left image shows the result after subtracting the original segment. The right side show the same, but the segment with the overlay was moved up by one voxel. The contours of the air gaps inside the phantom are reduced clearly, which confirms the assumption of the subsided phantom. The same technique was checked for overlay 3, but the result did not improve like in the case for overlay 4, so finally the technique was only applied for overlay 4.

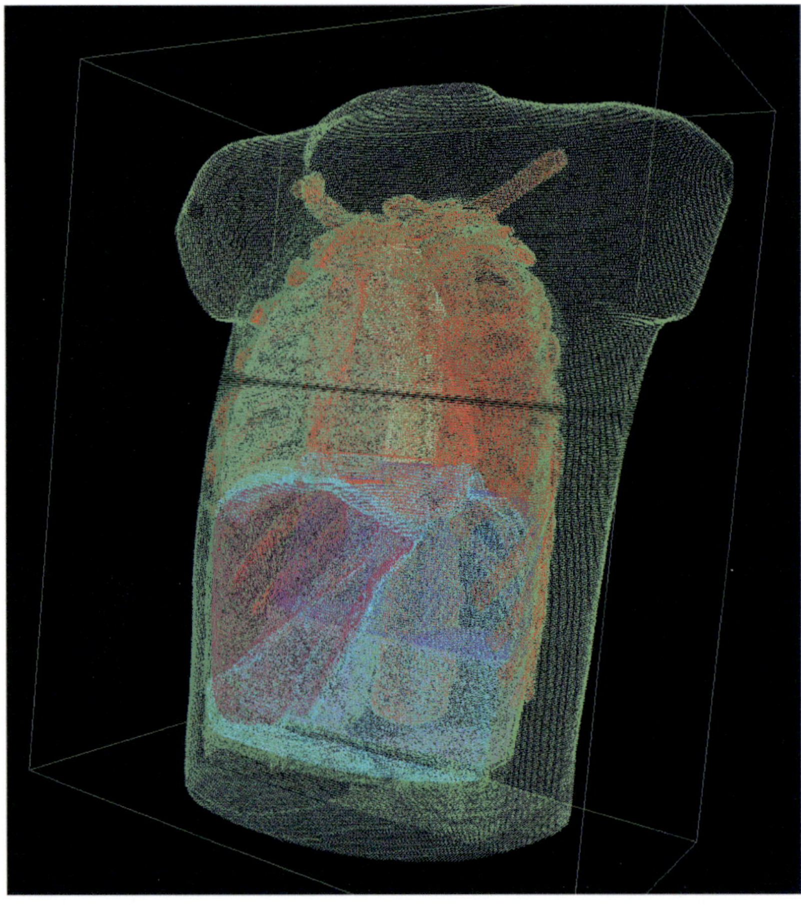

Figure 4.18: Finalized voxel model of the LLNL Torso Phantom. All organ surfaces are displayed as colored dots in a 3D-VRML world.

4.7.5   Lung Density Fluctuations

During the evaluation of the CT-scans a special characteristic of the lung lobes was noticed, which is most probable caused by manufacturing processes. The Hounsfield values of voxels at the lung lobe boundaries are considerably higher, i.e. showing a higher density, than the ones inside (figure 4.19). This was also noticed by others (Kramer *et al.*, 2000a). The denser areas at the boundaries could explain the tail of the peak in the histogram (figure 4.15). An explanation of the higher density at the lung lobe boundaries could be a special coating that was applied to prevent abrasion, which is particularly important for radioactive organs. The lack of gas bubbles at the boundaries is another evidence for this assumption. The effect of the denser boundaries may have an effect on validation experiments, because a uniform density distribution in the organs in the voxel model is assumed. This may have an effect on the radionuclide distribution on the radioactive lung lobes. The activity could be concentrated on the organ's boundaries. An experiment was performed to check if this effect plays a role. It is discussed further in section 6.2.4.

The foaming process of the polymerizing polyurethane is caused by carbon dioxide, which is formed by adding small amounts of water to the polymer starting material. When the lung lobes are poured into the molding, the formed CO_2 immediately produces pressure in the molding. So it needs to be tightly sealed right after pouring the polymer. This might be reason for loosing some mass and the weight differences observed. For the left lung lobes (ten pieces available at IVM, see tables 4.4 and 4.5) an average value of 532 g (standard deviation of 21 g) was found. The right lung lobes average was 692 g (standard deviation: 25 g). The casting of the solid organs seems to be less delicate. The livers (four pieces) seem to be much more constant and showed more an average of 2239 g (standard deviation: 17 g). To compensate the mass differences, the densities used in Monte Carlo simulations are calculated for every active organ separately.

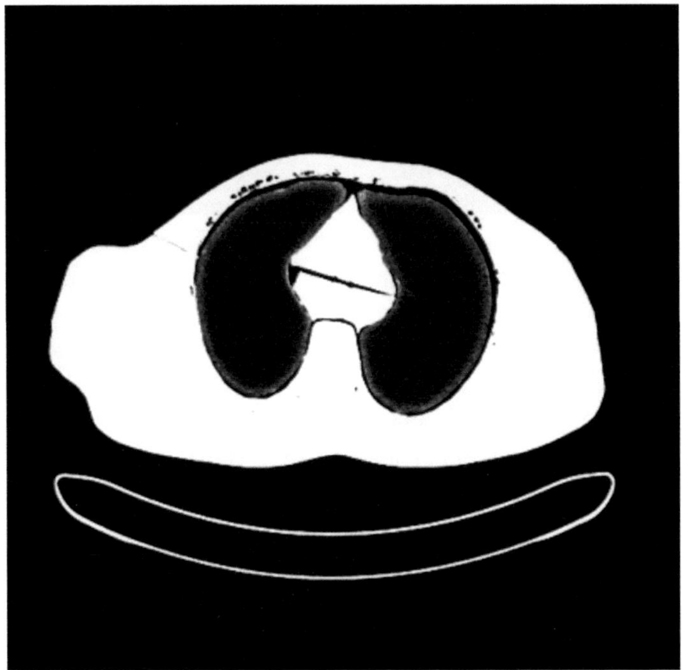

Figure 4.19: Image slice displaying the lung lobes rich in contrast. The inner of the lobes are clearly darker than the boundaries.

4.7.6 Volumes and Densities

With the methods described in this section a voxel model of the LLNL Torso Phantom has been created. Table 4.6 was produced from the final segmented voxel model with $512 \times 512 \times 633$ voxels. The segment's volumes were determined by multiplying the number of voxels of a segment with the voxel volume. The densities then have been calculated by dividing the weights (table 4.4) by the volumes. The bone's weight could not be determined as mentioned before, so a density value from the sample reports was taken to calculate their weights separately for the torso and its cover. The calculated values have been subtracted from the weighed values of the torso and its cover. An organ identification number has been assigned for all single segments.

At first glance, all the determined density values seem to be reasonable with two exceptions: the lung lobes and the overlays 1 and 2 of the A-set. The problem was to get clear specifications about this from the manufacturer RSD Phantoms, the successor of Humanoid Systems. The manufacturer did not provide such information most probable due to secrecy responsibilities.

Own investigations had been necessary to confirm the lung results, because ICRP suggested a value of 0.25 g/cm^3 (ICRP, 1975) for the lung. Later, Griffith *et al.* (1978) tried to produce tissue-equivalent lungs for the LLNL Torso Phantom prototype with a density close to that value. In the article, he set himself a range from 0.25 to 0.30 g/cm^3 for the lung density. Finally, the values obtained for our phantom were compared to the calculated value from the sample reports[22], which was 0.3211 g/cm^3. Also, another unpublished document[23] was found which confirmed the high value of 0.32 g/cm^3 relatively compared to the original ICRP-value. The average value of both lung lobes is 0.326 g/cm^3. The slightly higher value (less than 2%) compared to the sample report could be explained by larger air bubbles, which are excluded from the segment and have been defined as *air*. Nevertheless the method of keeping the mass constant and varying the density to compensate errors of the volume seems to give accurate results (see also section 6.2.6).

The calculated densities of most overlays seem to be dispersed between 1.044 and 1.085 g/cm^3. The overlays 1 and 2 of the A-set are more problematic, because their mass weighed, and hence their densities (1.143 and 1.206 g/cm^3) are much higher. A visual inspection did not show obvious differences to the overlays 1 and 2 of the other two sets. A possible explanation is that the manufacturer cannot reproduce the thickness of the overlays perfectly. A difference of about 1 mm in thickness can result in a mass fluctuation in the region of 100 g. Since there are not CT-scans of the A and C overlay sets, they have not been used in this work.

---

[22]Calculated from lung material sample report, part of the phantom's documentation (Radiology Support Devices, 1985)

[23]From *Instructions for mixing Plastinaut Lung Stock Material* by Alderson Research Laboratories, Stamford, CT, USA

Table 4.6: Volume and density of the phantoms segments. The air density was taken from dry air at 20°C and 101325 Pa.

Organ ID#	Part / Organ	Volume (cm^3)	Density (g/cm^3)
1	Torso	15010.0	1.091
2	Lower kidney block	1191.1	1.100
3	Upper kidney block	1759.8	1.086
4	Heart	1078.4	1.088
5	Liver envelope	2480.4	1.084
6	Liver	2038.4	1.085
7	Lymph node block	164.0	1.073
8	Left lung lobe	1563.9	0.337
9	Right lung lobe	2138.1	0.321
10	Torso cover	2877.7	1.040
11	Bones	1408.2	1.268
12	Overlay 1 (B)	1322.1	1.044
13	Overlay 2 (B)	2449.3	1.065
14	Overlay 3 (B)	3557.9	1.057
15	Overlay 4 (B)	4935.5	1.070
16	Air	84964.6	0.001204
17	Bones, torso cover	106.8	1.268
18	Bones, torso	1301.4	1.268
-	Torso cover w/ bones	2984.4	1.048
-	Torso w/ bones	16311.5	1.105
23	Overlay 1 (A)	1322.1	1.206
24	Overlay 2 (A)	2449.3	1.143
25	Overlay 3 (A)	3557.9	1.085
26	Overlay 4 (A)	4935.5	1.054
27	Overlay 1 (C)	1322.1	1.074
28	Overlay 2 (C)	2449.3	1.082
29	Overlay 3 (C)	3557.9	1.068
30	Overlay 4 (C)	4935.5	1.070

4.7.7 Coarsening

Large voxel models are highly memory consuming. MCNP5 does not have subroutines, which can handle large voxel models quickly during the initialization of the input file. The MCNP5 code was manipulated by Wang *et al.* (2005) and enabled the calculation of transmission photon radiographs of a portion of the VIP-Man with just 100 million voxels (Xu *et al.*, 2000). A developer of MCNP5 was coauthor and performed the manipulations, which were quite complicated and needed full knowledge of the code. More advanced for voxel models is MCNPX 2.5.0. The initialization takes only seconds, while MCNP5 takes minutes. But still MCNPX can handle just a limited number of voxels. If MCNPX is compiled in the 32-bit version, the code will only handle memory up to $2^{32} = 4$ Gigabytes. Not only the voxel model array with one-byte-per-voxel occupies the memory, but also auxiliary arrays associated to the voxel model consume considerable amounts of memory.

The newly created voxel model is composed of $512 \times 512 \times 633 \approx 166 \times 10^6$ voxels. Superfluous air voxel can be cut from the model with Voxel2MCNP automatically. An algorithm checks all six sides of the model for voxel layers only composed of air voxel. The result was a model $512 \times 285 \times 628 \approx 92 \times 10^6$ voxels.

Still the number of voxel was too high for the 32-bit version of MCNPX. That is the reason why a coarsening algorithm was developed for the voxel model of the LLNL Torso Phantom that merges a cubic voxel cluster to one big new voxel, by determining the mode of the cluster's voxel values. Taking the mean value instead would lead to absurd results, because there is no definition for non-integer values. In case of bi- or multimodal results a random generator chooses one of most abundant values, each with equal probability. If inevitable, the cluster cubes were filled with air voxels to avoid indivisible residuals at the borders of the model.

The applicability of this algorithm was checked by comparing the volumes and center of gravity of the organs before and after the coarsening. A cube of $3 \times 3 \times 3 = 27$ voxels has been merged to one new voxel resulting in a model with a new voxel edge length of three times the old edge length. The coarsening algorithm reduced the number of voxels to approximately 3.4×10^6. This model size can be managed more easily with the available hardware. Table 4.7 shows the effect of the coarsening algorithm on the organ volumes. The coarsening was repeated several times due to the random component of the algorithm and an average value was taken for the volumes. It was observed that the random component caused volume fluctuations less than 0.5%. Organs with a relatively large surface area lose volume during the coarsening algorithm while organs with small surface areas gain volumes. This effect and its impact on detector efficiency will discussed in section 6.2.2. Ideally the organ volumes stay the same, which was almost the case in this example. The volumes changed from -0.9% to $+0.7\%$. The bones, which is the organ with the largest surface area to volume ratio, lost 0.9%, while other organs gained in volume. The values of the coordinates for the center of mass of the organs varied from -1.3 mm to -0.3 mm.

It should be mentioned that changes in the shape of the organs were not considered. It cannot be excluded that such changes in shape may have an effect on the simulations. Considerable changes of the shape - except more rough surfaces of the organs compared to the original resolution - are not expected and could not be observed visually. As a result, the coarsened 3^3-cluster model was preferred for future simulations with the model of the LLNL Torso Phantom, since it keeps important properties of the original model in acceptable ranges.

Table 4.7: Changes in volume for the reduced model (coarsened in 3^3-voxel cluster) compared to the original model.

ID #	Organ	Original volume (cm^3)	3^3-model volume (cm^3)	Relative volume of 3^3-model to original-model (%)
1	Torso	15010	15032	100.1
2	Lower kidney block	1191	1195	100.4
3	Upper kidney block	1760	1767	100.4
4	Heart	1078	1080	100.1
5	Liver envelope	2480	2492	100.5
6	Liver	2038	2042	100.2
7	Lymph node block	164	165	100.7
8	Left lung lobe	1564	1569	100.4
9	Right lung lobe	2138	2146	100.4
10	Torso cover	2878	2891	100.5
11	Bones	1408	1396	99.1
16	Air	85421	85352	99.9

4.7.8 Geometric Validation

The new model of the LLNL Torso Phantom was checked if it has the same proportions and measures like the real phantom. This was done to verify that the CT-scanner delivered valid information about the voxel dimensions.

Landmarks have been defined for points of the phantom that are easy to recognize and to define in both, the model and the phantom. The landmarks (figure 4.20) of the phantom have been measured with a new developed positioning system on the phoswich detector rack introduced in section 6.3.1, while the phantom was reclined on the examination table of the PBC on a stiff wooden board to prevent tilting. Voxel2MCNP's 3D-viewer (section 4.6.2, see also figures 4.7 and 6.27) was used to determine the coordinates of the corresponding landmarks on the model. To do this, the model was placed with the same alignment on the virtual examination table in the model of the PBC.

A translational transform has been used to superimpose both sets of coordinates. The transform parameters were determined by calculating the mean deviation for each coordinate. Table 4.8 shows the results. The standard deviation of the deviations for each coordinate was calculated printed in the table and gives evidence for a successful geometric validation of the new developed model of the LLNL Torso Phantom. No specific trend of the data was observed that would indicate a rotational shift in any direction. Considering errors of the positioning system, errors from locating the landmarks virtually with Voxel2MCNP due to voxelization, and of course errors due to the decision of the user how to locate the landmark precisely at both systems, the geometry and the dimension of the model can be assumed to be valid.

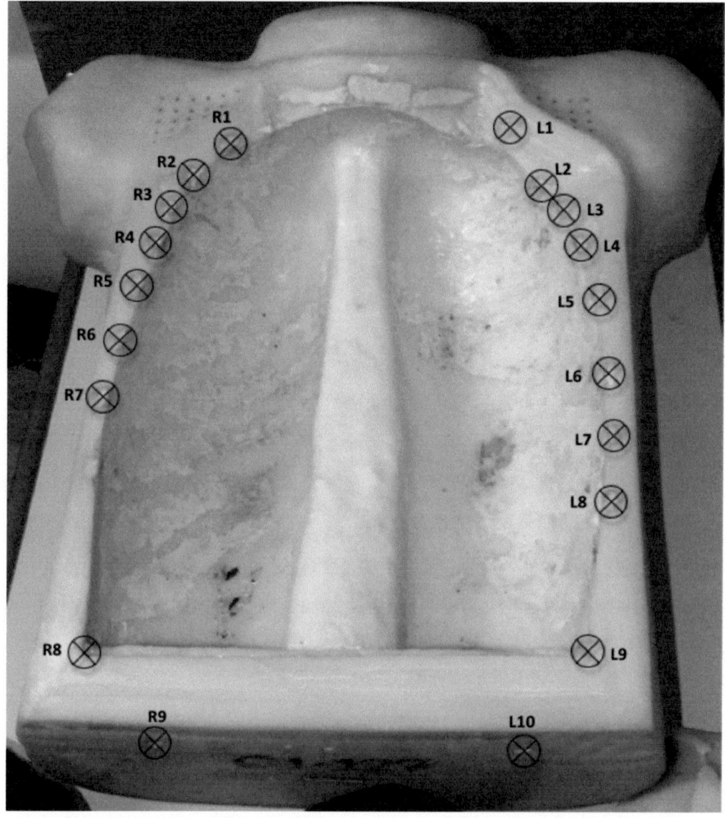

Figure 4.20: Landmarks (marked with ⊗) defined on the LLNL Torso Phantom (without chest cover and inner parts). The substitute material representing the ribs (R1 to R7, L1 to L8) on the chest's edges can be well distinguished from the rest of the phantom and can be used for landmark definitions as well as corners (R8, L9) and holes (R9, R10).

Table 4.8: Landmark coordinates of voxel model and the deviations to the real phantom after a translation.

Landmark	Coordinates of voxel model			Deviations to real phantom after translation		
ID#	x (mm)	y (mm)	z (mm)	x (mm)	y (mm)	z (mm)
1R	1640.15	813.75	1733.15	-1.50	-0.06	3.32
2R	1617.25	810.00	1758.75	-1.90	-0.81	4.52
3R	1608.25	811.90	1786.00	-1.10	-0.91	3.67
4R	1602.50	810.95	1813.00	-0.65	-0.86	2.67
5R	1597.00	812.90	1847.25	-0.25	-1.91	2.02
6R	1591.75	813.75	1887.50	-1.00	-1.06	3.37
7R	1587.75	812.80	1925.00	-0.90	0.99	2.17
8R	1599.40	803.50	2068.00	1.45	-0.31	-0.63
9R	1634.50	773.00	2101.60	2.95	-1.81	-1.03
1L	1814.05	810.00	1716.80	-3.50	3.19	0.17
2L	1833.30	791.25	1750.00	-1.95	1.44	0.87
3L	1845.00	795.00	1773.60	-0.55	1.19	-0.93
4L	1853.00	799.70	1805.25	-1.15	-0.11	-0.98
5L	1859.55	805.30	1847.60	-0.40	1.49	-1.23
6L	1862.40	806.25	1901.60	1.25	0.44	-2.23
7L	1861.50	805.25	1940.00	2.45	-0.56	-2.73
8L	1857.65	804.30	1978.80	2.60	-1.51	-4.13
9L	1839.40	804.50	2061.60	1.05	2.69	-4.53
10L	1808.90	772.25	2097.60	3.05	-1.56	-4.33
			Mean	0.00	0.00	0.00
			Std. dev.	1.87	1.49	2.85

Chapter 5

Validation

Monte Carlo methods enable us to model scenarios in the virtual world that we are not able to set up in the real world. Monte Carlo codes like MCNPX can be used for the calculation of the detector efficiency, given a specific in vivo measurement scenario. However, in order to apply Monte Carlo methods, a proper validation of the models used is required.

High-energy point source scenarios can be straightforward to validate. However, scenarios with low-energy photon emitters are challenging. The probability that low-energy photons interact with matter is high. Even small layers of solid material can attenuate them considerably. Radioactive volume sources like organs, in which radionuclides are incorporated, are far more complex to simulate than simple point sources.

The detector models need to be precise as many scientists (Schläger, 2007; Liye *et al.*, 2007) realize when they come to a point to compare the results from real measurements with the ones from simulations. Technical drawings from the detector manufacturer often do not represent the as-built status and are not accurate.

Also, challenging is the validation of simulations with voxel phantoms, because both the original phantom and its voxel model are required at the same time. Most popular voxel models for example NORMAN (Jones, 1996), Zubal's torso model (Zubal *et al.*, 1994), VIP-Man (Xu *et al.*, 2000) (see also section 4.5.2) are not available to in vivo laboratories for real measurements. Sometimes the original was destroyed in the process of producing the digital model.

Since Monte Carlo methods are becoming more popular, members of EURADOS[1] organized an intercomparison (Gómez-Ros *et al.*, 2008) between different laboratories. A knee-phantom was loaded with ^{241}Am and measured with germanium detectors in one laboratory. For the other laboratories enough information, including the voxel model of the knee phantom, have been provided to enable all participants to simulate the scenario. The ISF also participated in this intercomparison successfully.

Encouraged by this positive result the methods and tools (chapter 4) developed and used in this thesis needed to be validated with the detectors in the in vivo counting facilities at the IVM. This was done by performing MCNPX simulations and comparing them to real measurements.

The first short section gives a review of the validation work done on the WBC, followed by the presentation of the validation results of the phoswich detectors of the PBC. Finally, since more complex and realistic validation investigations became possible, the last section of this chapter describes the validation of measurement scenarios with HPGe detectors with the newly created voxel model of the LLNL Torso Phantom (see also sections 3.2 and 4.7).

5.1   Whole Body Counter

The WBC was simulated with the IGOR phantom and the BfS-bottle phantom models (see section 3.2, page 20). This work was performed and described by Sessler (2007) during his diploma project. This section gives a short overview on this study. Performed simulations are based on the methods introduced in chapter 4.

5.1.1   IGOR

Instead of using a voxel model, Sessler (2007) developed combinations of boxes and cylinders to represent the mentioned phantoms. Figure 5.1 shows the measurement setup of IGOR P_4 in the WBC. The corresponding MCNPX model can be seen in figure 5.2.

---

[1] European Radiation Dosimetry Group

Figure 5.1: The IGOR P_4 phantom in the WBC. VRML representation of the model of IGOR P_4 in the WBC with particles tracks computed from PTRAC data extracted by Voxel2MCNP.

Table 5.1 compares the data from simulations with data from measurements with different IGOR configurations and different radionuclides. Statistical counting errors have been kept below 1% for all simulations and real measurements in the ROI. The relative deviations were determined between 0.5% (P_2 with ^{133}Ba) and 9% (P_6 with ^{60}Co). The deviations were not found to be dependent on nuclide or energy, respectively. The simulated efficiencies slightly overestimate the measured ones, but overall, the results show a good agreement.

One possible explanation for the higher simulated efficiencies could be a not-fully developed detector model. Since the WBC detectors and their electronics are old, the detectors might have lost their full efficiency during all the years of service. The crystals are hygroscopic and might have drawn water. Wet crystal parts cannot contribute to the scintillation processes anymore. Another consideration is that the detector drawings might not be as built and therefore, the models for the simulation could be based on wrong assumptions. Furthermore, the activities of the used radionuclide sources in the experiment have uncertainties of $\pm 5\%$. Another possible error source could result from

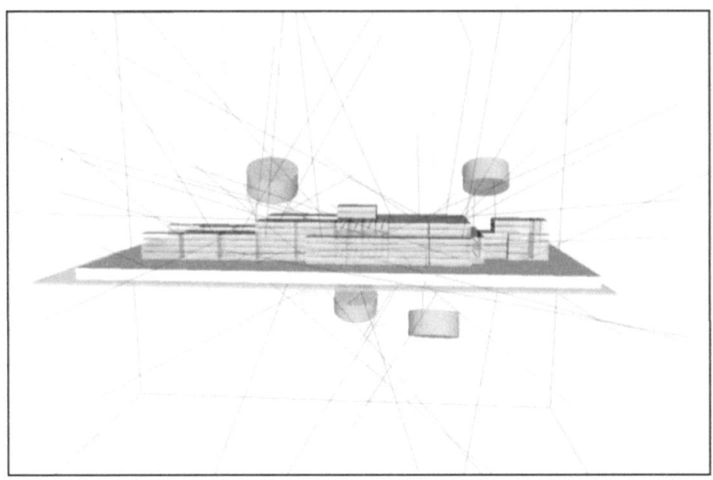

Figure 5.2: VRML representation of the model of IGOR P_4 in the WBC with particles tracks computed from PTRAC data extracted by Voxel2MCNP.

positioning. As a result, Sessler (2007) showed that moving or shifting the phantom model by just a few centimeters can influence the resulting efficiencies by a few percent. At the moment test persons cannot be positioned as exact it was done in the experiments with the lifeless phantoms. This means, the experimental accuracy shown by Sessler's simulations cannot be applied in routine anyway. Considering all possible errors, the simulation data in table 5.1 can be accepted as valid.

Table 5.1: Comparison of detector efficiencies of the four NaI(Tl) detectors of the WBC from simulations and from real measurements in the corresponding ROIs of the radionuclides ^{133}Ba, ^{137}Cs, ^{60}Co, and ^{40}K using the IGOR phantom in configurations P_1 to P_6 (Sessler, 2007).

IGOR phantom		Σ detector efficiencies of WBC-detectors (counts/decay)			
Configuration ID	Mass (kg)	^{133}Ba Simulation	Measurement	^{137}Cs Simulation	Measurement
P_1	12	2.03×10^{-2}	n/a	1.66×10^{-2}	n/a
P_2	24	1.92×10^{-2}	1.92×10^{-2}	1.58×10^{-2}	1.57×10^{-2}
P_3	50	1.63×10^{-2}	1.57×10^{-2}	1.36×10^{-2}	1.34×10^{-2}
P_4	70	1.41×10^{-2}	1.34×10^{-2}	1.19×10^{-2}	1.16×10^{-2}
P_5	90	1.31×10^{-2}	1.27×10^{-2}	1.11×10^{-2}	n/a
P_6	110	1.20×10^{-2}	1.13×10^{-2}	1.03×10^{-2}	n/a
		^{60}Co simulation	measurement	^{40}K simulation	measurement
P_1	12	1.35×10^{-2}	n/a	1.24×10^{-2}	n/a
P_2	24	1.30×10^{-2}	n/a	1.19×10^{-2}	n/a
P_3	50	1.13×10^{-2}	1.10×10^{-2}	1.05×10^{-2}	1.02×10^{-2}
P_4	70	1.00×10^{-2}	9.50×10^{-3}	9.40×10^{-3}	8.80×10^{-3}
P_5	90	9.40×10^{-3}	8.90×10^{-3}	8.80×10^{-3}	8.50×10^{-3}
P_6	110	8.80×10^{-3}	8.10×10^{-3}	8.30×10^{-3}	8.10×10^{-3}

5.1.2   Bottle Phantom

Similar results had been produced for the BfS-bottle phantom (figure 5.3), which will not be discussed in detail in this thesis. As an extract for the work done, ^{40}K-spectra are plotted in figure 5.4 which show a good match between the simulated and measured spectra in the peak region. The statistical fluctuations in the measured spectrum below 1.3 MeV is due to the short measurement time. More obvious is the issue below 0.2 MeV have already been observed by others (Breustedt, 2004) and can not be fully explained at this stage. Originally, ^{40}K was modeled as a pure photon emitter, but the majority of the emitted particles are β^--particles, hence the measurements observed in this low energy range might be a result of Bremsstrahlung. Simulations with β^--particles have been repeated to consider this. A normalized β^--spectrum was added to the photon spectrum. The results improved slightly, but the low energy range remains underestimated in the simulation. However, the WBC results are not influenced by those findings, as there are no ROIs in the low energy range. Theoretically speaking the simulations could be improved, but as the accuracy of the simulations in the ROIs in the energy range above 200 keV is sufficient for the application of numerical efficiency calibration methods described in chapter 4, an improvement was not necessary for the experiments conducted for this thesis.

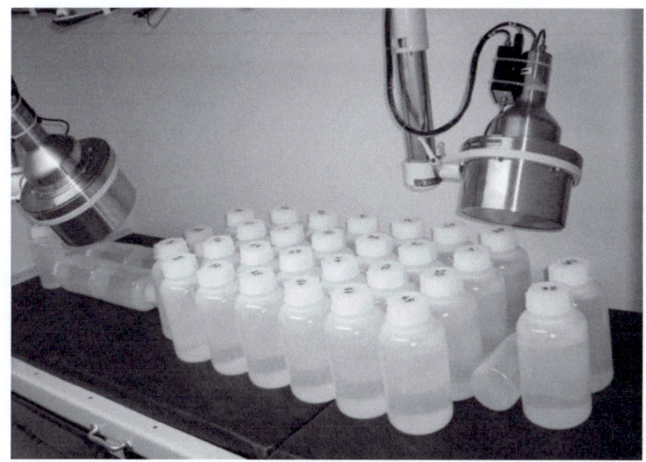

Figure 5.3: Bottle phantom (70-kg-configuration) in the WBC.

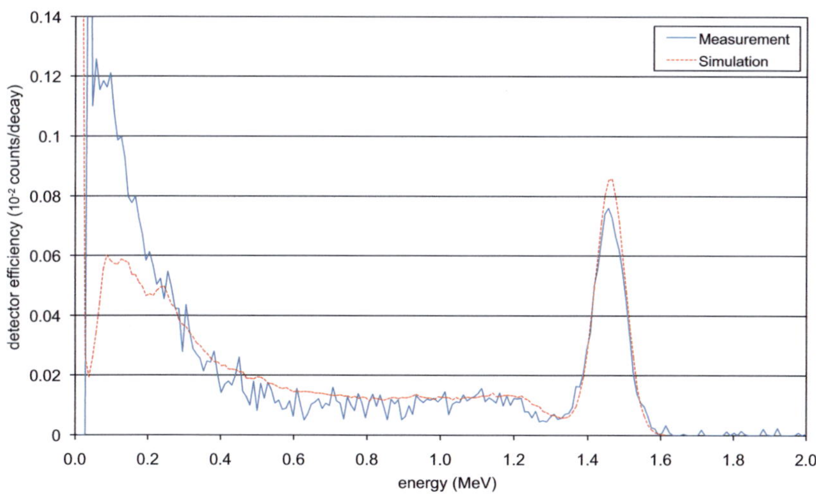

Figure 5.4: Spectrum of the bottle phantom (70-kg-configuration) loaded with ^{40}K in the WBC.

5.2 Phoswich Detector Model Validation

The phoswich detector pair in the PBC at IVM (see section 2.1) has a great detector efficiency due to its large detector diameter, but the energy resolution is comparable poor to the observed resolutions of germanium semiconductor detectors. This disadvantage was acceptable, as the phoswich pair has proven to be the reliable detector pair of choice for over thirty years of service.

Over the years the detectors have changed and their energy resolution has decreased. This is most likely due to hydration of the hygroscopic detector crystals. But it is at present not clear to which extent this could have an effect on detector efficiency. Modeling such aged detectors is a difficult task as also the electronics of the detector system must be considered. The photomultiplier tube, the amplifiers, and the anti-coincidence circuits contribute to the distinct shape of the spectrum. A systematic approach was used to generate an acceptable model of the phoswich detectors that is able to resemble measured spectra. Part of this work will be published in the upcoming annual report (Leone and Hegenbart, 2009) of the Central Safety Department, which is the organization the ISF originated from.

5.2.1 Experiments for Spectral Shape

First of all, point sources have been measured with the detector to gather spectral data that helps to model the detectors. The point sources have been attached to the main column of the Phoswich rack by a string. The phoswiches were not rotated vertically, but inclined by $45°$, symmetrically pointing to the point source on the string, i.e. the source was placed at the intersection of both detectors' axes and the main axis of the phoswich rack. The rack was moved to a position in the middle of the PBC far away from scattering objects like the examination table. The detector position and the source position was recorded and later used for set up all the simulations accordingly.

Following radionuclides have been used: ^{241}Am, ^{133}Ba, ^{22}Na, ^{137}Cs, ^{60}Co, ^{57}Co, ^{210}Pb, and ^{237}Np. The measurement times ranged from 100 s to 1000 s, resulting in a statistical counting error in the ROIs of less than 1% and a smooth, low-noise spectrum. The dead time of the detectors was not higher than 5% to avoid an overload of the detector system that cause erroneous results. For the short measurement time of the point sources, the contribution from background radiation was negligible. Nevertheless, the weekly recorded background spectrum was scaled to the measurement time and subtracted from the measured point source spectra.

5.2.2 Experiments for Reproducibility

A second series of measurements was performed for which the point sources were put on a steel table. A reference position was defined with the point source vertical below the main column. The position of table, source and detector was recorded and applied to all simulations. After one measurement, the reference setup was changed and put back in place to check the reproducibility of this positions. Additionally, the detector rack was moved in three directions (x-, y-, and z-direction) up to 5 cm from the reference position. Point sources of four radionuclides, namely ^{241}Am, ^{22}Na, ^{137}Cs, and ^{60}Co were used. The measurement times were 300 s, resulting in a statistical counting error in the ROIs of less than 0.5% and a smooth low-noise spectrum shape. Once again, the dead time of the detectors was not higher than 5% and also here, the background spectrum was normalized and subtracted from the point source spectra.

5.2.3 Density Determination of the Aluminum Oxide Layer

The density of the aluminum oxide (Al_2O_3) powder layer behind the beryllium entry window of the phoswich detector is a sensitive parameter as the layer is used as a light reflector to keep scintillator light inside the detector (Doerfel *et al.*, 2006).

Available technical drawings do not give information on the thickness of this layer. The thickness may vary due to the manufacturing process, i.e. it depends how hard the crystal was pressed into its bed of Al_2O_3. To simplify the problem, the thickness can be assumed to be constant leaving the density as the only parameter that needs to be determined experimentally. Literature reports density value for aluminum oxide powder ranging from 1.17 g/cm^3 for loosely poured powder up to the solid state with a density of 3.94 g/cm^3 (Ortwin Rave, 2009; Merck, 2009).

For ^{241}Am, the density has a high influence on 17.5keV peak and the low energy area below 35 keV, while the influence is comparable low in the energy range from 40 to 90 keV. The density of the layer has negligible effects on measurements of the CsI(Tl)-crystal that covers the high-energy range.

Doerfel *et al.* (2006) assumed a value of 1 g/cm^3 for the density and a thickness of 1 mm. This value as well as the GEB parameters (see also section 4.3 and equation 2.2) were taken from Doerfel *et al.* (2006) as a starting point for the model. Simulations with these values could not reproduce the measured spectra sufficiently. The low-energy ranges were overestimated with this value.

To solve this problem, two ROIs have been defined. The first range was defined from 10 to 35 keV and the second from 40 to 90 keV. The simulated and the measured spectra were normalized to the unit *counts/decay* for this investigation (see equation 4.1). Ideally, the number of counts in the ROI of the real measurement are equal to those from the simulation. The phoswich detectors always showed fewer counts than in the simulation, possibly due to idealized assumptions of the model. The ratio of the corresponding ROIs (real measurement / simulation) were calculated. Then, the density was varied in the simulations until the ratios of both ROIs, the low and the upper energy range, had the same ratio.

The result was a density of 1.9 g/cm^3 for phoswich detector #1 and 2.2 g/cm^3 for detector #2.

5.2.4   Determination of FWHM

The peaks in the spectra are assumed to have a Gaussian shape with the mean at μ and a standard deviation of σ. FWHM is defined as peak width of a Gaussian peak at the half maximum (equation 2.3). The resolution of a detector is expressed by the FWHM as a function of the energy (see equation 2.2). The photo-peaks at the energies given in table 5.2 have been fitted with Gaussians $f_g(E)$ according to equation 5.1. The height of the peaks was adjusted by factor n_g.

$$f_g(E, n_g, \mu, \sigma) = n_g \cdot e^{-(E-\mu)^2/(2\sigma^2)} \qquad (5.1)$$

Table 5.2: Photo-peak energies rounded to full *keV* for the radionuclides investigated for the FWHM determination of the phoswich detectors.

Low-energy range (NaI(Tl)-det.)		High-energy range (CsI(Tl)-det.)	
Nuclide	Peak energy (keV)	Nuclide	Peak energy (keV)
Am-241	18	Na-22	511
Ba-133	30	Cs-137	661
Pb-210	46	Co-60	1173
Am-241	60	Na-22	1274
Ba-133	81	Co-60	1332
Np-237	86	K-40	1460
Co-57	122		

It was tried to separate the peaks of interest from other neighboring peaks. The major difficulty concerning the nuclides detected with the NaI(Tl)-crystal was to differentiate the photo-peak from the other peaks at different energies, i.e. the peak at 59.5 keV of the ^{241}Am from the X-rays at lower energy. The first step for low-energy spectra was to obtain the net peak areas of the peaks. An algorithm (Canberra Industries, Inc., 2009) was used, which provides the subtraction of the continuum underlying the peak. The continuum originates from background radiation or/and Compton scattering from other higher-energy peaks above the peak of interest. Details on the algorithm can be found in pages 300–303 in the software manual (Canberra Industries, Inc., 2006). All further calculation were performed with corrected peak spectra.

For the high-energy photon emitters, it was tried to separate the photo-peak at energy E_p from the Compton shoulder. In a Compton interaction, the energy transferred E_T from a photon to an electron depends on the deflection angle θ of the scattered photon (equation 5.2). The maximum energy is transferred at an angle of $\theta = \pi$. The so-called Compton edge at E_C (equation 5.3) marks this maximum energy transfer in a spectrum.

$$E_T(\theta) = E_p - \frac{E_p}{1 + \frac{(1-\cos\theta)E_p}{m_e c^2}} \tag{5.2}$$

$$E_C = E_T(\pi) = E_{T,max} = \frac{2E_p^2}{m_e c^2 + 2E_p} \tag{5.3}$$

with

$$m_e c^2 = \text{electron's mass times light speed squared} \approx 0.511 \text{ MeV}$$

The Compton shoulder is mainly relevant for higher energy photo-peaks, because E_C is within the peak width and contributes to the peak underground. Furthermore the probability of Compton interaction for low-energy photons is low compared high-energy photons. According to the work of Baccaro *et al.* (1997) the Compton shoulder can be described with a Fermi-Dirac distribution $f_f(E)$ (equation 5.4). The factor n_f scales the height of the distribution and factor K determines the shape of the slope.

$$f_f(E, n_f, K) = \frac{n_f}{1 + e^{(E-E_C)/K}} \tag{5.4}$$

The photo-peak was fitted using both, a Gaussian and a Fermi-Dirac distribution simultaneously in case of higher energy peaks. In the cases of overlapping peaks, more Gaussians (and Fermi-Dirac distributions) were used for fitting the spectra. Gnuplot's fit function (Williams *et al.*, 2009) was used to fit a theoretical formula composed of Gaussians and Fermi-Dirac distributions with the measurement data. Free parameters were the scaling factors n_g and n_f for both distribution types, the mean μ and standard deviation σ for Gaussians, the factor K for the Fermi-Dirac distribution. Initial starting parameter values for the fit function were guessed. The parameters were not constrained or related to each other for simplification. Just the Compton edge energy E_C was fixed and calculated by equation 5.3 from the theoretical photo-peak energy E_p.

An example is the identification of the triple peak of ^{237}Np with a main peak at 86.5 keV (see figure 5.5). Here, three Gaussians $f_{g,1} + f_{g,2} + f_{g,3}$ were fitted. Fermi-Dirac distributions can be neglected for such low energies. Another example is the separation of the peaks of ^{60}Co at 1.173 and 1.333 MeV that are overlapping due to the poor resolution of the CsI(Tl) detectors. They were fitted by two Gaussians and two Fermi-Dirac distributions: $(f_{g,1} + k) + f_{f,1} + f_{g,2} + f_{f,2}$ (see figure 5.6). The constant k was introduced to take into account that the left peak sits on top of the continuum of the right peak.

With equation 2.3, the FWHM of each photo-peak in table 5.2 was calculated from σ. The obtained value was plotted and fitted for each detector with equation 2.2, using the Gnuplot fit function (see figures 5.7 and 5.8). The parameters a and b and their asymptotic standard errors are shown in table 5.3. Leaving the optional correction parameter c free for the fit resulted in high errors. Therefore, c was set to zero since, just as well as Doerfel *et al.* (2006) did.

The calculated fit functions for the FWHM-function of the NaI(Tl)-detectors lie within the points' error bars. The fits for the CsI(Tl)-detectors are comparable poor. The reason, why the fit functions miss most of the points' error bars, is unknown. Also Doerfel *et al.* (2006) encountered problems fitting these functions with equation 2.2. A possible explanation for this could be unknown electronic effects or physical detector deficiencies. Further investigations are necessary to explain this observation.

Table 5.3: Determined parameters a and b for the detector resolution according to equation 2.2.

Detector ID#	Crystal type	a (MeV)	b (MeV$^{0.5}$)
5	NaI(Tl)	-0.013 ± 0.001	0.144 ± 0.006
7	NaI(Tl)	-0.0090 ± 0.0007	0.099 ± 0.003
6	CsI(Tl)	-0.049 ± 0.009	0.18 ± 0.01
8	CsI(Tl)	-0.016 ± 0.01	0.12 ± 0.02

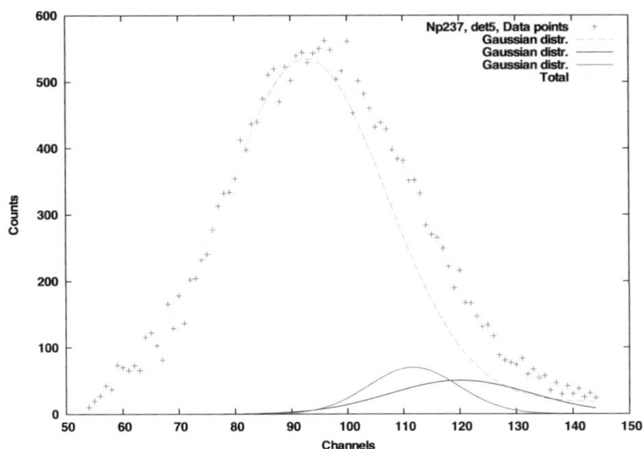

Figure 5.5: The plot shows the result of fitting the measured ^{237}Np peak triplet with three Gaussians with Gnuplot's fit function.

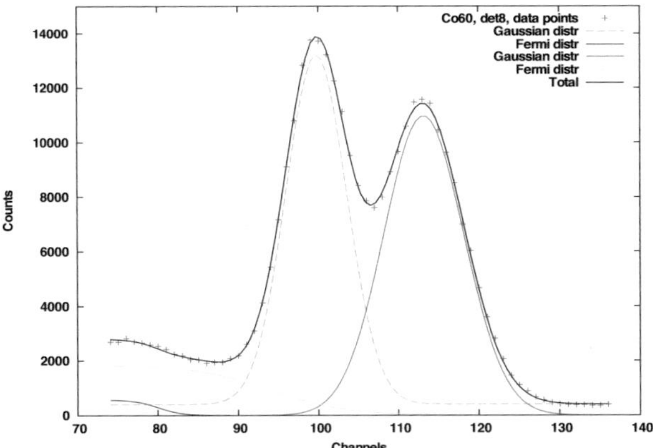

Figure 5.6: The plot shows the result of fitting the measured ^{60}Co peak doublet with two Gaussians and two Fermi-Dirac distributions with Gnuplot's fit function.

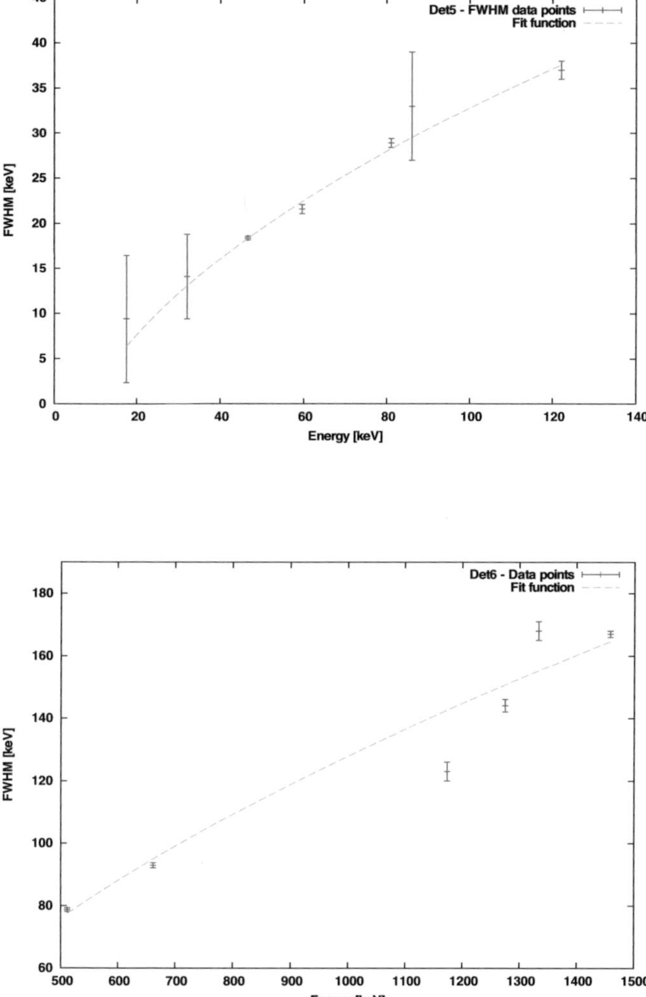

Figure 5.7: Two plots showing the fits for the energy dependent FWHM function for each detector of the phoswich 1. The upper plot represents the fit functions for NaI(Tl)-detector 5 and the lower plot represents the one for CsI(Tl)-detector 6. Error bars show the asymptotic standard errors of each peak fit.

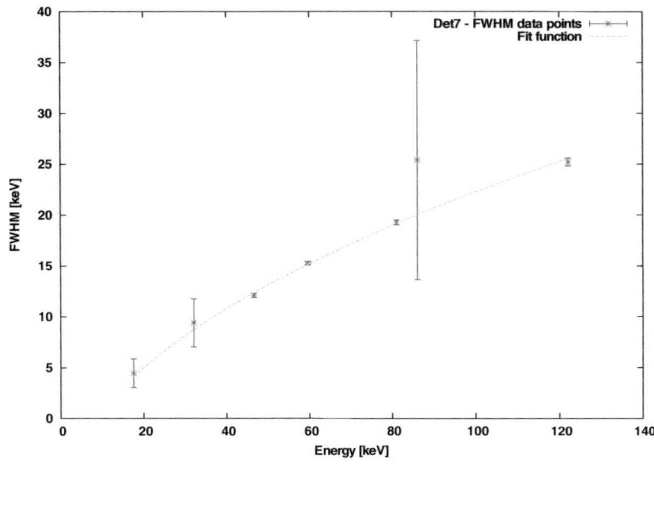

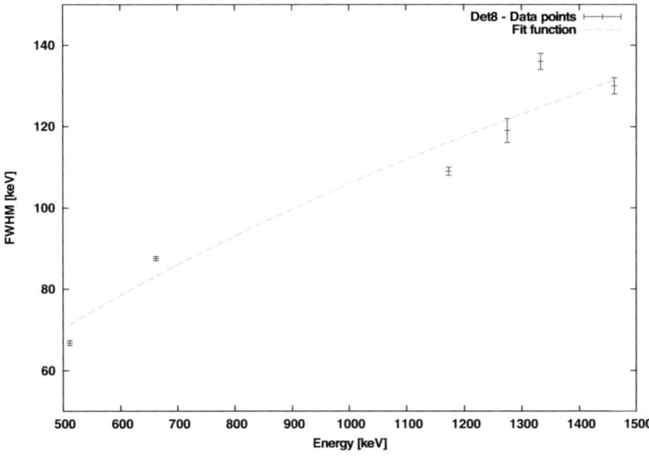

Figure 5.8: Two plots showing the fits for the energy dependent FWHM function for each detector of the phoswich 2. The upper plot represents the fit functions for NaI(Tl)-detector 7 and the lower plot represents the one for CsI(Tl)-detector 8. Error bars show the asymptotic standard errors of each peak fit.

5.2.5 Validation Results

Tables 5.4 and 5.5 show the ratios of the simulated and measured efficiencies. The ratios are not equal to one, because ideal detectors are assumed in the simulation. The simulated efficiencies are always higher probably due to the fact, that the real detectors lost their initial intrinsic efficiency due to hydration. Especially the NaI(Tl) detector crystal #7 seems to have a low yield.

Nevertheless, the ratios are constant and reproducible for each crystal in the same ROI within small fluctuations of a few percent. The fluctuations are acceptable considering the statistical counting errors for the Monte Carlo simulation and the real measurements as well as the errors of the source activity. According to the point source certificates, the uncertainties (σ) of the activity ranged from $\pm 1.9\%$ for ^{60}Co up to $\pm 4\%$ for ^{241}Am.

The re-positioning of the detector rack in the *reference* position showed no effect on detector efficiency ratios (see table 5.4). Also the shifts of the detector rack up to 5 cm in all directions starting from the reference position could be reproduced in simulations (tables 5.4 and 5.5) and showed no trends.

The determined yields were applied to the spectra. Figures 5.9 and 5.10 exemplarily show the spectra comparison of measurements and simulations for one NaI(Tl)- and one CsI(Tl)-detector. The shape of the measured spectra is reproduced well by the simulated spectra. Only in the low energy range below 20 keV for the NaI(Tl)-detectors, the broadening of the X-ray peaks is not reproduced perfectly. The fit of the FWHM function could not be extended confidentially to low energies, which is just a few *keV* above the detector's energy-detection limit. The theoretical energy-dependent FWHM formula (equation 2.2) might in this case not be applicable in practice. Nevertheless, for ^{241}Am, the obtained fit function is sufficient to reproduce the spectrum in the range from 20 to 80 keV, which is the most important energy range for this radionuclide.

Table 5.4: Detector efficiency ratios (measurements/simulations) for the four phoswich detector-crystals for the radionuclides ^{241}Am and ^{137}Cs within their specific ROIs.

| | Detector efficiency ratios measurement / simulation | | | |
| | ^{241}Am, 0.04–0.08 MeV | | ^{137}Cs, 0.574–0.749 MeV | |
Shift (cm)	det. 5	det. 7	det. 6	det. 8
reference	0.94	0.86	0.91	0.97
ref. repetition 1	0.95	0.87	0.91	0.98
ref. repetition 2	0.94	0.87	0.91	0.97
x+5	0.96	0.87	0.91	0.99
x-5	0.94	0.84	0.90	0.97
y+1.7	0.93	0.86	0.90	0.98
y-5	0.94	0.85	0.91	0.96
z+5	0.96	0.88	0.92	0.98
z-5	0.91	0.85	0.90	0.97
mean	0.941	0.861	0.907	0.974
std. dev.	0.016	0.014	0.006	0.009

5.2.6 Conclusion

The density of the Al_2O_3 layer right in front of the NaI(Tl)-detectors could be modeled successfully. For the determination of the FWHM, some peak fits can be improved by e.g. constraining the parameters to theoretical values. The energy-dependent relationship of the FWHM for the CsI(Tl)-crystal requires further investigation, e.g. of the electronics and possible detector deficiencies.

Nevertheless, the application of the obtained parameters in the simulations shows satisfactory agreements with the measured spectra. Except for some problems in the lowest measurable energy range of the NaI(Tl)-detectors, the spectral shape could be reproduced satisfactorily. The problems in this low energy range could originate from imperfect cross section data that is used for Monte Carlo simulation. This guess needs further examination, although there are references from studies with different Monte Carlo codes, different detectors, and different measurement scenarios that also encountered problems reproducing the spectra satisfactorily in this energy range (Gómez-Ros *et al.*, 2008).

Table 5.5: Detector efficiency ratios (measurements/simulations) for the four phoswich detector-crystals for the radionuclides ^{60}Co and ^{22}Na within their specific ROIs.

| | Detector efficiency ratios measurement / simulation | | | |
| | ^{22}Na, 1100-1465 MeV | | ^{60}Co, 1056–1458 MeV | |
Shift (cm)	det. 6	det. 8	det. 6	det. 8
reference	0.94	0.97	0.96	0.96
x+5	0.96	0.95	0.96	0.96
x-5	0.92	0.91	0.95	0.95
y+1.7	0.94	0.97	0.97	0.97
y-5	0.97	0.96	0.95	0.96
z+5	0.96	0.95	0.97	0.96
z-5	0.93	0.93	0.95	0.93
mean	0.945	0.950	0.960	0.958
std. dev.	0.017	0.021	0.010	0.014

The detector efficiencies reproduced by the simulations are valid considering a constant yield. Consistent results were found for different detector positions.

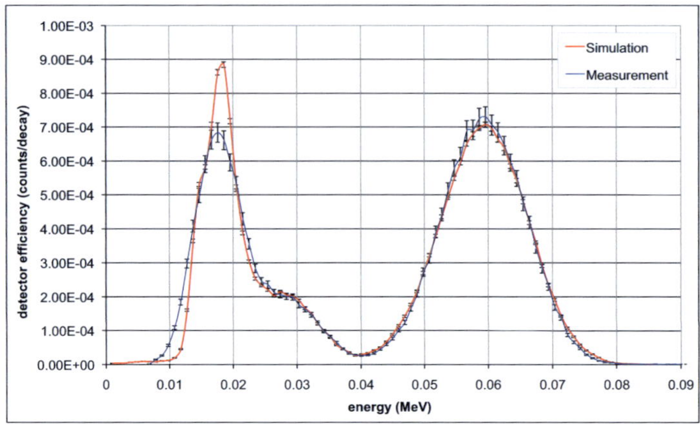

Figure 5.9: Measurement of an ^{241}Am point source in the reference position with NaI(Tl)-detector 7 compared with the corresponding simulated spectrum corrected by the corresponding yield from table 5.4. Error bars showing the statistical uncertainties for simulation and measurement. Additionally, the uncertainty of the point source's activity is included in the error bars for the measurement.

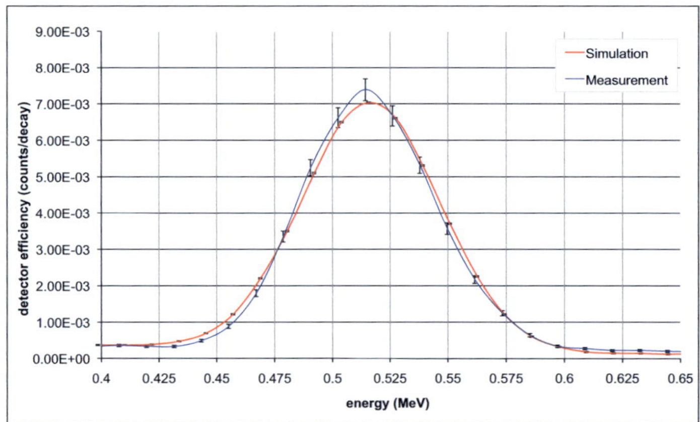

Figure 5.10: Measurement of an ^{22}Na point source in the reference position with CsI(Tl)-detector 8 compared with the corresponding simulated spectrum corrected by the corresponding yield from table 5.5. Error bars showing the statistical uncertainties for simulation and measurement. Additionally, the uncertainty of the point source's activity is included in the error bars for the measurement.

5.3 HPGe-Detector Scenario

Typical in-vivo measurements have been performed with the LLNL Torso Phantom. Parts of the investigation presented in this section have been published (Hegenbart *et al.*, 2009).

The detector used was one of the new XtRa HPGe-detectors (section 2.1). It has a high energy resolution compared to scintillation crystal detectors. This detector was chosen to validate simulations, since its model (section 4.4.3) was well validated with point-source experiments before by Marzocchi (2008) at IVM. The positioning, the source organ, the radionuclide, and the chest wall thickness were varied to check for associated systematic errors.

Liver and lung have been used as source organs. The chest wall thickness was varied by overlays on the phantom. Two low-energy photon emitting radionuclides, ^{241}Am and ^{239}Pu, were used. The experiments revealed the positioning as the crucial part of the experimental setup, which is discussed in detail in section 6.1.1.

5.3.1 Setup

The detector was mounted on a in-house constructed, height adjustable frame (see figure 5.11) with the detector crystal axis pointing perpendicular (parallel to the y-direction). A laser rangefinder (Bosch DLE 50) and a ruler (for distances smaller 15 cm) was used to determine the position of the detector and the phantom in the PBC chamber according to the defined coordinate system. The phantom was positioned on the floor, its position determined by landmarks (section 4.7.8). The simulation geometries were set up for the virtual model accordingly.

The measurements were performed starting with reference setups in which the detector was centered just above surface of the phantom over the organ of interest (figure 5.12). Concentric markings on the torso cover helped to define these centers. The detector position was then adjusted by changing one coordinate per measurement: the x-coordinates from −6 cm to +6 cm, z-coordinates (from right to left) from −4 cm to +6 cm, and the y-coordinates (postero-anterior direction, i.e. from the torso's back to the front) from 0 cm to +6 cm. For the y-axis, the

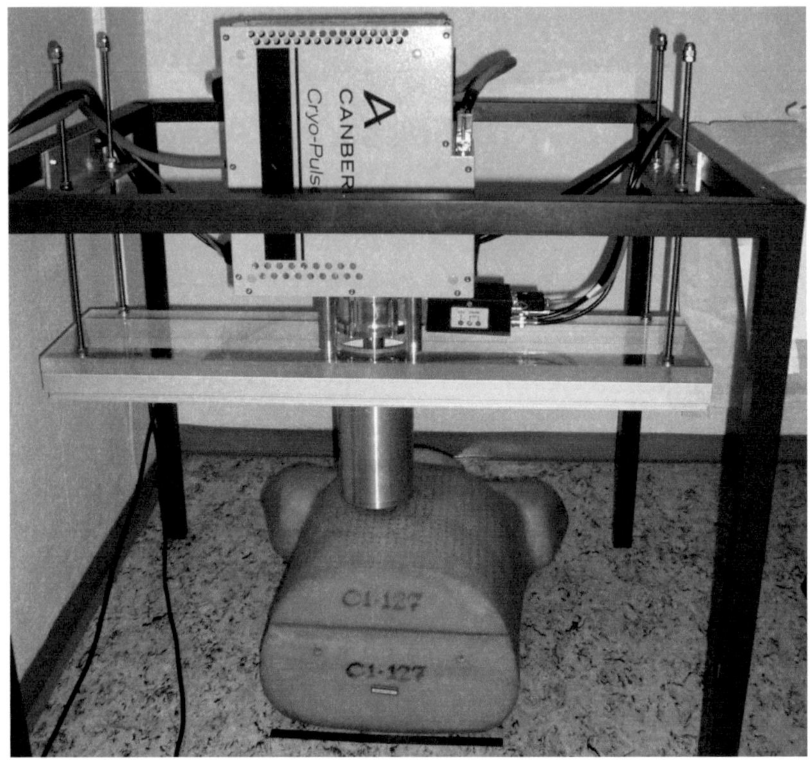

Figure 5.11: Detector mounted in the frame with the LLNL Torso Phantom underneath.

adjustments were only possible in the positive range, since the detector cannot be moved closer towards the phantom.

The results were evaluated in Microsoft Excel with algorithms taken from Genie 2000 (Canberra Industries, Inc., 2006). The chamber's background was determined overnight prior to the actual experiments and subtracted from the measured results.

Monte Carlo simulations were performed using MCNPX with default physics features in `mode e p`. The MCNPX input file was created with Voxel2MCNP (section 4.6). For the simulations it was assumed that the radionuclides were distributed homogeneously throughout the organ volume. It was assumed that the radioactive organs are equal in volume

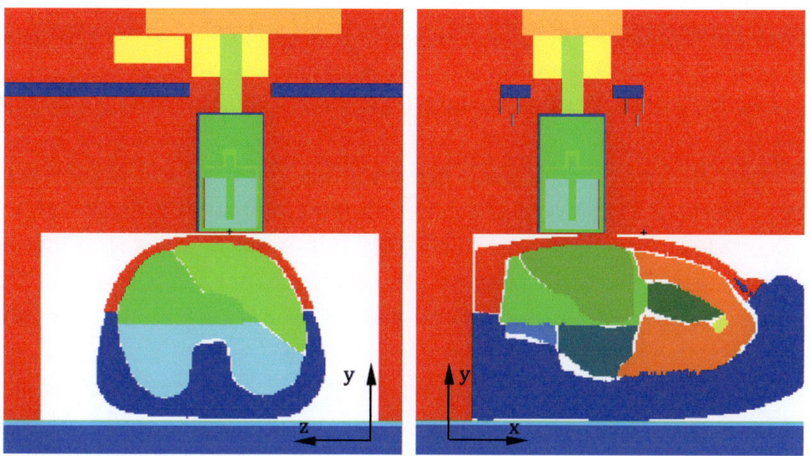

Figure 5.12: Screenshot from MCNPX's 2D-plotter shows two sections of liver reference setups with the XtRa HPGe-detector model on top of the voxel model of the LLNL Torso Phantom (Hegenbart *et al.*, 2009).

to the non-active ones and the densities in the simulations were adjusted to compensate observed weight differences (table 4.5), especially for the lung (section 4.7.5). Source radionuclides from Voxel2MCNP's nuclide library (section 4.6.1) were used. For the various simulations the particle histories ranged from 10×10^6 to 200×10^6, depending on the desired statistical counting error within the ROI. The ROI was set to be six times the FWHM to cover the complete peak. The relative statistical counting error (equation 3.3) was always kept below 1%.

5.3.2 Results and Discussion

Overall the comparisons of the simulations and the measurements showed a good agreement. Figure 5.13 shows a comparison of a simulated and a measured spectra. The spectra match well, except for the peak broadening in the simulated peaks below 30 keV that could not be reproduced perfectly. However the main peak for ^{241}Am at 59.5 keV, which is important to calculate the detector efficiency, is reproduced appropriately.

For a more detailed evaluation the ratio of the two spectra from figure 5.13 was determined (see figure 5.14). The plot of the ratios shows almost a straight line at ratio 1 from 30 to 60 keV, but besides the peak broadening at lower energies two further issues are visible.

The first issue is in an additional 2-10% overestimation of the net area in the energy range from 47 to 52 keV, compared to the net area of the main 59.5 keV peak. Such behavior has already been observed by other authors (Gómez-Ros *et al.*, 2008), who compared different Monte Carlo codes. It is therefore likely that this issue is not dependent on the simulations performed or on the specific experimental setup. The observed behavior may have been caused by the underlying transport algorithms used by MCNPX.

The second issue is the fluctuation of the ratio between simulated and measured spectra for the main ^{241}Am peak at around 60 keV. This issue is related to different factors, such as imperfect parameters of the Gaussian energy broadening and a small offset of the main energy peak between experimental setup and MCNPX input files caused by energy calibration. However, these issues in the spectrum were not investigated further, as they do not affect the measurement in the ROI.

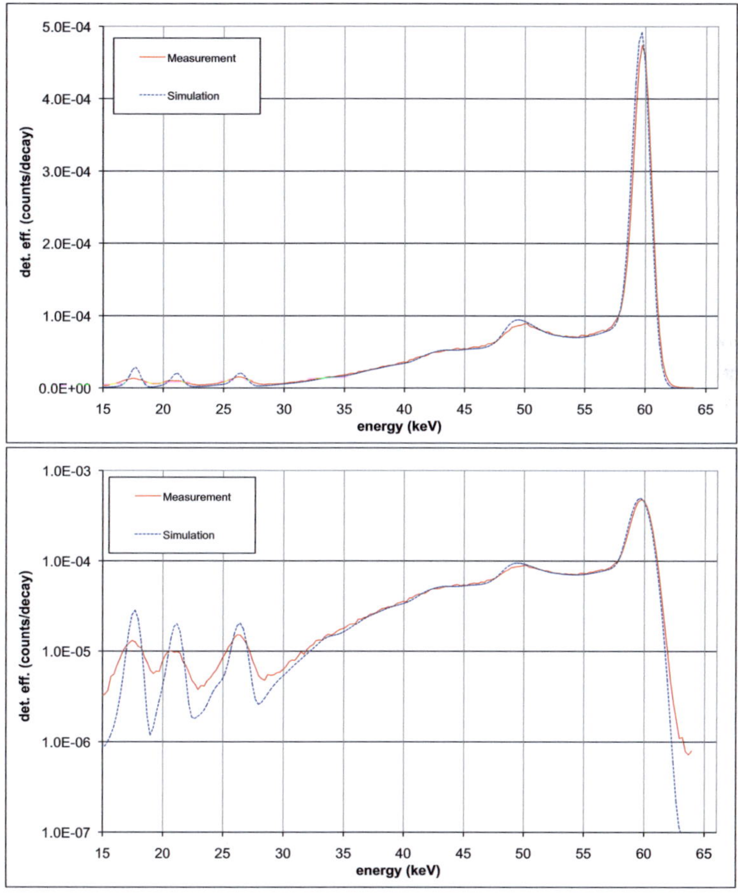

Figure 5.13: Measured and simulated ^{241}Am spectra of the liver determined with the XtRa HPGe-detector and its model in a reference position with linear (upper plot) and logarithmic (lower plot) ordinate. The background spectrum was subtracted from the measured spectrum. The spectra presented are representative for all other recorded ^{241}Am spectra.

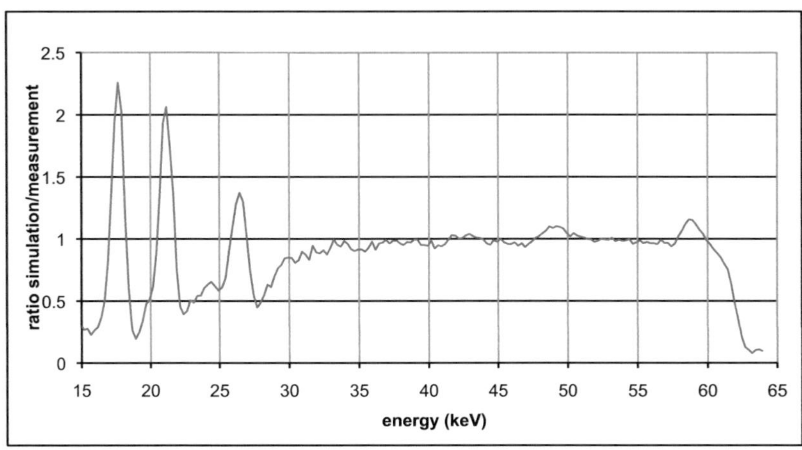

Figure 5.14: Ratio of measured and simulated detector efficiencies of the ^{241}Am spectra from figure 5.13. This plot is representative for all other recorded ^{241}Am spectra.

Table 5.6 shows the detector efficiencies of the performed measurements and simulations with ^{214}Am as the source radionuclide. The shifts in z-direction above the liver show remarkable detector efficiency ratios of between 0.99 to 1.02. The y-direction ratios vary from 0.93 to 0.97, which is also close to 1.00. Greater fluctuation have been observed for the shifts in x-direction. Here the ratios changed from 0.99 to 1.09. The high ratios of the $(x - 6$ cm)- and $(x - 4$ cm)-shift could be caused by positioning errors (see figure 5.15).

Positioning errors of ± 0.5 cm have been assumed for the positioning of the detector relative to the phantom. They were evaluated by simulations, where the position was shifted by this amount in the respective direction. Errors caused by positioning in other directions were not considered. The shifts were applied only in the virtual model as the important parameter is the relative shift detector-phantom, not the actual positioning. The results were displayed as error bars in figure 5.15.

Applying different overlays show detector efficiency ratios ranging from 0.95 to 0.98.

Table 5.6: Detector efficiencies within the ^{241}Am ROI (57 to 64 keV) in different configurations. Three reference positions have been defined A, B and C. The values for the reference positions A and B differ slightly due to a disassembly and reassembly of the experiment that took place on different days. Respective offsets are displayed in centimeters.

Organ	Overlay ID#	Position (offset from ref. (cm))	Detector efficiency (counts/decay) Simulation	Measurement	Ratio sim./meas.
Liver	0	x–6	2.27×10^{-3}	2.09×10^{-3}	1.09
Liver	0	x–4	3.06×10^{-3}	2.91×10^{-3}	1.05
Liver	0	x–2	3.89×10^{-3}	3.82×10^{-3}	1.02
Liver	0	ref. A	4.32×10^{-3}	4.26×10^{-3}	1.01
Liver	0	x+2	4.43×10^{-3}	4.38×10^{-3}	1.01
Liver	0	x+4	4.31×10^{-3}	4.31×10^{-3}	1.00
Liver	0	x+6	3.85×10^{-3}	3.90×10^{-3}	0.99
Liver	0	z–4	3.49×10^{-3}	3.50×10^{-3}	0.99
Liver	0	z–2	3.94×10^{-3}	3.91×10^{-3}	1.01
Liver	0	ref. A	4.32×10^{-3}	4.26×10^{-3}	1.01
Liver	0	z+2	4.15×10^{-3}	4.08×10^{-3}	1.02
Liver	0	z+4	3.74×10^{-3}	3.71×10^{-3}	1.01
Liver	0	z+6	3.10×10^{-3}	3.10×10^{-3}	1.00
Liver	0	ref. B	4.22×10^{-3}	4.36×10^{-3}	0.97
Liver	0	y+1	3.69×10^{-3}	3.86×10^{-3}	0.95
Liver	0	y+2	3.23×10^{-3}	3.44×10^{-3}	0.94
Liver	0	y+3	2.84×10^{-3}	3.03×10^{-3}	0.94
Liver	0	y+4	2.51×10^{-3}	2.67×10^{-3}	0.94
Liver	0	y+6	1.98×10^{-3}	2.13×10^{-3}	0.93
Liver	0	ref. B	4.22×10^{-3}	4.36×10^{-3}	0.97
Liver	1	y+0.6	3.41×10^{-3}	3.47×10^{-3}	0.98
Liver	2	y+1.2	2.75×10^{-3}	2.89×10^{-3}	0.95
Lung	0	ref. C	1.88×10^{-3}	2.10×10^{-3}	0.89
Lung	1	y+0.6	1.52×10^{-3}	1.79×10^{-3}	0.85
Lung	2	y+1.2	1.32×10^{-3}	1.51×10^{-3}	0.87

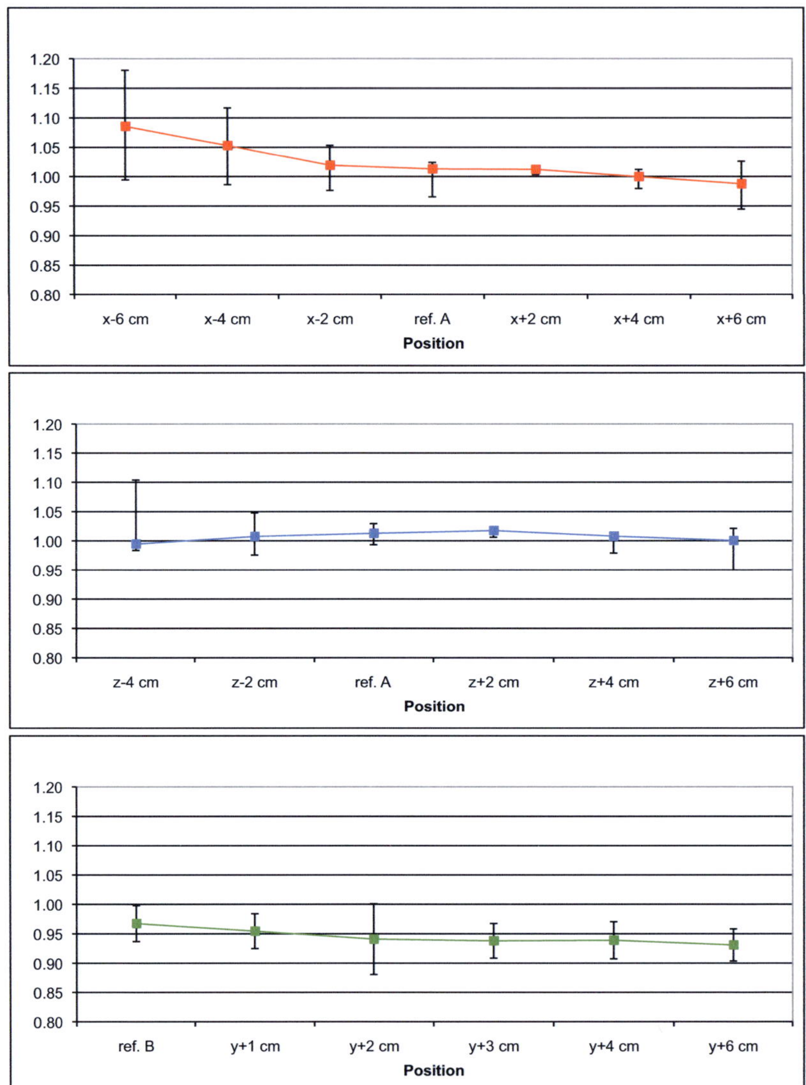

Figure 5.15: Ratios of measured and simulated detector efficiencies in different configurations (x-, y-, and z-shifts). The error bars indicate positioning errors. They represent shifts of ± 0.5 cm in the respective direction and have been estimated with MCNPX simulations.

The measurements and simulations performed with the lung lobes and different overlays also showed constant ratios. The observed ratios are lower compared to those of the liver. This can have several reasons. The source organs' activity is specified with an error of $\pm 5\%$ and might explain the slightly larger deviation of the ratios from 1. A non-uniform density distribution of the phantom's lungs and probably also the associated non-uniform distribution of the activity may contribute to this (see also section 6.2.4).

A problem was discovered during the evaluation of the measurements with the ^{239}Pu organs, as impurities of ^{241}Am were found in the organs. The organs have been manufactured in the 1980s. ^{239}Pu is artificially created from irradiating uranium with neutrons. Various reactions are taking place at the irradiation and various isotopes of plutonium are produced. Chemically, ^{239}Pu cannot be separated from ^{241}Pu (half-life of 14 years), only centrifuges are able to do this. Almost two half-lifes passed and a considerable amount of the ^{241}Pu impurities decayed to ^{241}Am according to equation 5.5.

$$^{241}Pu \rightarrow {}^{241}Am^+ + e^- + \bar{\nu}_e \tag{5.5}$$

Figure 5.16 shows the measured spectrum of the liver. It also shows an attempt of summing up weighed spectra from ^{241}Am and ^{239}Pu to obtain the measured spectrum. Approximately 0.5% of the total activity is assumed to come from ^{241}Am. Assuming an age of 27 years, the initial molar percentage of ^{241}Pu was about 0.013%. Still the spectra do not match appropriately, which may be due to other impurities. It was therefore decided not to investigate the results from the ^{239}Pu organs further.

5.3.3 Conclusion

Partial body counting measurements of the LLNL Torso Phantom with varying parameters like detector position, low energy photon emitting radionuclides, source organs and chest plate overlays have been compared to Monte Carlo simulations performed with the new voxel model of the LLNL torso phantom and with a precise model of a Canberra XtRa

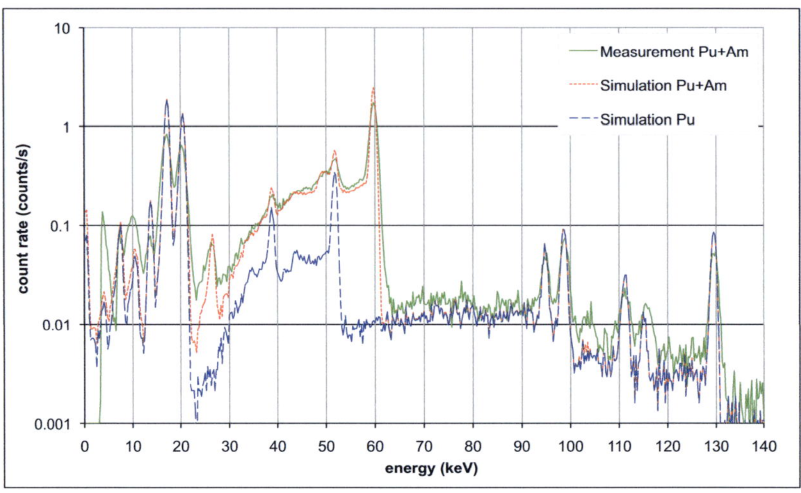

Figure 5.16: Measured and simulated ^{239}Pu spectra with ^{241}Am impurities as well as a ^{239}Pu-only spectrum.

detector in the virtual measurement environment of the PBC at IVM introduced in chapter 4.

Besides some small discrepancies, the measured and the simulated spectra overall show a good agreement. The ratios of the detector efficiencies of the measurements and the simulations are within acceptable range around 1.

The experiments to validate the used simulation environment succeeded for the ^{241}Am organs. The ratios are ranging from 0.93 to 1.09 for the liver and 0.85 to 0.89 for the lung, which is acceptable considering the errors mentioned.

A high degree of sensitivity of the system to the relative phantom-detector placement was observed. This will be investigated in detail in section 6.1.1. Impurities have been found in the ^{239}Pu organs of the LLNL Torso Phantom that must be considered in the calibration measurements. The agreement of simulations and corresponding measurements and the stability of the ratios leads to the conclusion that the simulation environment has no considerable systematic errors.

Chapter 6

Sensitivity Analyses

In this chapter the effects of different parameters on detector efficiency are investigated. The investigations have been performed virtually by Monte Carlo simulations, since the simulation environment introduced in chapter 4 has been found to be a valid representation of the real measurements (chapter 5). Simulations have the advantage, that setting specific parameters can be done precisely and reproducible.

The detector system of choice was the phoswich detector pair, which is the *workhorse* of the PBC at the IVM. The advantage of its high detector efficiency overcomes the drawback of its poor energy resolution. Therefore the focus in this chapter lies on the phoswich detectors.

For the first section of this chapter, the model of the phoswich detector system was used to determine the effects of various geometric parameters on the detector efficiency in different measurement scenarios. For the second section special parameters beyond geometry are investigated on their effect on detector efficiency. The third section summarizes the results and describes concluded measures that have been implemented to reduce systematic errors for numerical and experimental detector efficiency calibration.

6.1 Geometry Parameters

6.1.1 Detector Position in Relation to a Subject

The first parameter to be checked was the position of the detectors in relation to the subject. The phoswich rack has a number of different detector positioning options. The rack is attached to a post that runs on rails which are attached to the ceiling of the PBC chamber. It can be moved in in x- and z-direction as described in figure 6.1. The height (y-direction) of the apparatus can be changed with the use of a motor. The rack itself can be rotated vertically around the axis of the post. The rack holds two detectors at either end, each of them can be rotated vertically and horizontally. The rack will be described in more detail in this chapter in section 6.3.1.

Setup

The Voxel2MCNP software (section 4.6) was used to place the MEET Man in $(4\ mm)^3$ voxels onto the examination table of the PBC. The longitudinal axis of the examination table as well as the MEET Man were parallel to the z-axis. The lung was chosen as the source organ using ^{241}Am as the radionuclide. Phoswich detector 1 positioned over the left lung lobe was used in a reference position according to the standard lung counting position, i.e. 25° rotated and 25° inclined parallel to the chest surface. The distance between detector front and MEET Man was about 1 cm. This was enough space for the virtual detector movements (see figure 6.2). The movements in x- and y-direction were from −1 cm to +1 cm, while in z-direction they were from −2 cm to +2 cm. The angles of the horizontal and vertical axes were varied from −10° to +10° from the reference position. The number of simulated particle histories was 10^7. The statistical counting error was kept below 0.3% (equation 3.3) in the ROI of 20 to 80 keV throughout this experiment.

Results and Discussion

Table 6.1 presents the results of this investigation. The rotations and inclinations had very little effect on the detector efficiency. Observed

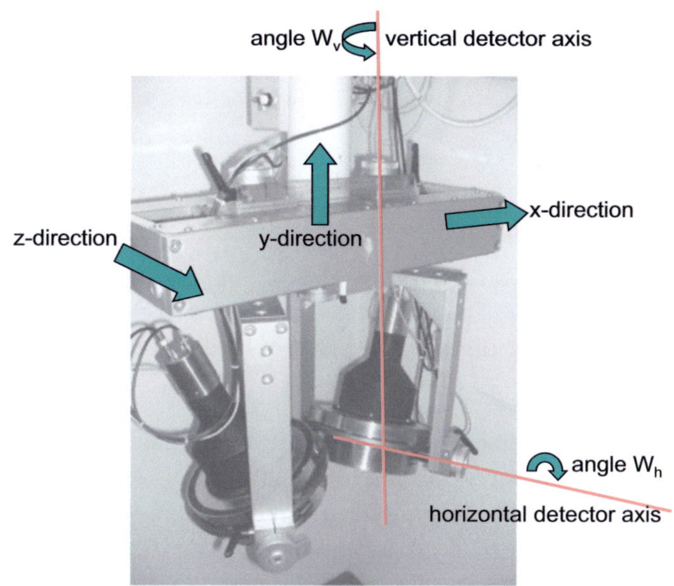

Figure 6.1: The picture shows the standard rotation and inclination (both 25°) of the phoswich detectors 1 and 2 used for lung counting. Furthermore it shows the directions in which the phoswich detector rack can be moved. The main axis of the post of the rack was never turned for experiments in this work, so it remained at the 0°-position parallel to the x-axis of the room (see also section 6.3.1)

deviations when turning the detector by 5° were in the range of the statistical counting error. This was different for the translations. The detector efficiency changed by –4.4% to +2.0% for –2 to +2 cm movement in z-direction. The efficiency for x- and y-movements changed by +4.0% to –5.2% and +6.3% to –5.9% for –1 to +1 cm, respectively. The results also show an obvious improvement of the detector efficiency for shifts in negative x-direction. In additional simulations, the second, right lung lobe was also defined as a source and the detector was moved above the middle of both lobes, which resulted in a higher yield. The second phoswich that is usually positioned mirror-symmetrical to the y-z-plane to partially compensate the changes caused by the x-movements. This was checked by investigations not presented in this work.

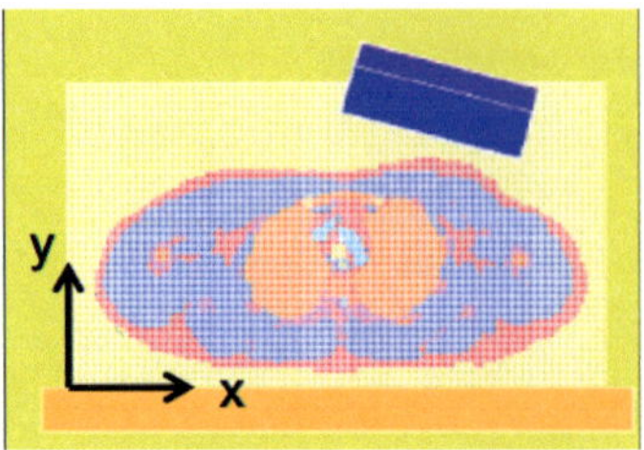
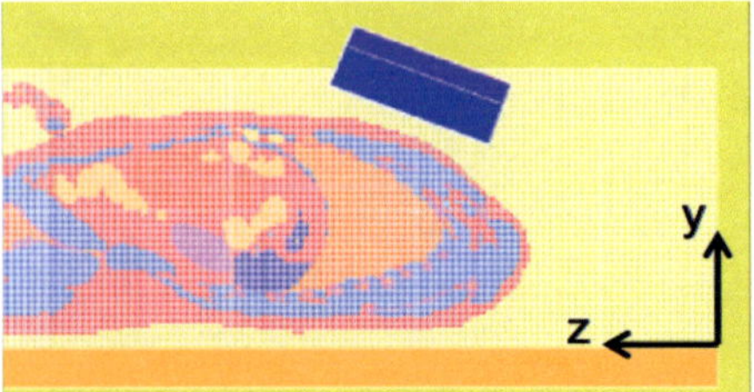

Figure 6.2: A transversal and a sagittal section plotted by MCNPX's 2D-plotter show the reference position of the phoswich detector 1 (blue color) over the right lung lobe (light orange color) of MEET Man voxel model.

The effects of the y-movements are no surprise either. The attenuation of additional layers of air according to equation A.4 is one reason, but to a greater extend this effect originates from the inverse-square law. For point sources, i.e. spherical photon fields, the intensity of radiation is reduced by the square of the distance. However, in this case a volume source is used, hence it is hard to calculate the attenuation analytically. That is why Monte Carlo methods have been applied.

It was found that the relationship of y-displacement with detector efficiency in the small range of 4 cm can be assumed as linear. Performing a linear regression for the x- and y-displacements resulted in a R^2-value (squared Pearson Correlation Coefficient) of greater than 0.997. The slope for the x-displacement was calculated as -1.06×10^{-4}counts·photon^{-1}mm^{-1} and -1.28×10^{-4}counts·photon^{-1}mm^{-1} for y-displacement.

Conclusion

While rotations and inclinations have little a effect, translations, especially in y- and x-direction of a few centimeters results in considerable changes of the detector efficiency.

6.1.2   Voxel-Wise Displacements of the Source Organ

The following investigations have been performed to check the influence of voxel-wise organ erosion and displacement.

Setup

In this setup the MEET Man voxel model with $(4\ mm)^3$ voxels was used. The boundary lung voxels were eroded by one voxel layer and replaced by water voxels to substitute the liberated tissue resulting in the lung lobes only being surrounded by one voxel space. The lung was moved in increments of plus/minus one voxel in all three directions. Figure 6.3 illustrates the procedure. Furthermore, the two lung lobes were also moved towards and away from each other in x-direction (left-right direction, see figure 6.4 for orientation). Finally, the already eroded lung was eroded for a second time by one voxel layer to examine the effect of erosion more intensively.

For each individual modification the detector efficiency for a typical ^{241}Am-lung counting measurement with the phoswich detectors was checked. The detectors were positioned rotated and inclined by $25°$, hence placed in the standard position for lung counting about 1 mm above the voxel model. This position was kept for all modifications. The number of simulated particle histories was 10^7, resulting in a statistical counting error for the sum of both phoswich detectors of less than 0.2% in the ROI (20 to 80 keV).

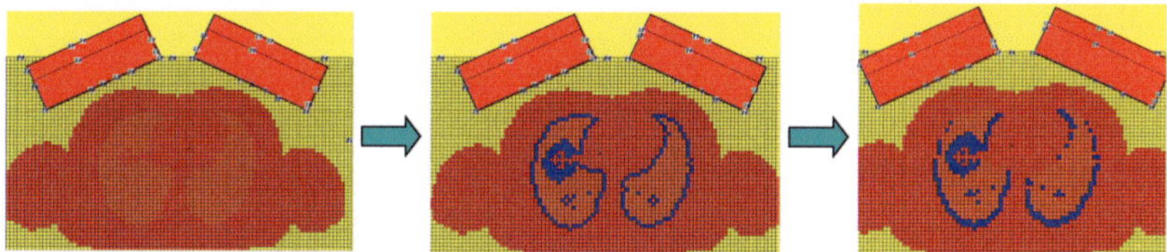

Figure 6.3: Sections showing the procedure of eroding lung lobes (left image) by one voxel layer (middle image) and anterior displacement towards the detectors (right image).

Results and Discussion

The erosion of the lung and hence the reduction of the lung volume seem to result in lower detector efficiencies. Since the density of water voxels is roughly four times higher than for lung voxels, the erosion also represents an increase of mass between detector and lung. The method

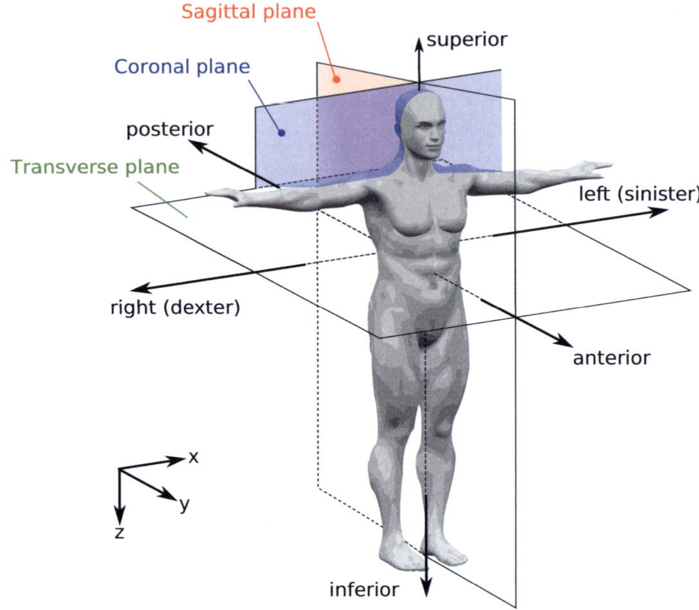

Figure 6.4: Human anatomy planes modified from Wikipedia's Human_anatomy_planes.svg.

used in this investigation is based on erosion of boundary voxels. The problem here is that the center of gravity is moving with every erosion, since the lung lobes are not point-symmetrical objects. The different centers of gravity of the eroded lungs make a comparison difficult. The effect of erosion on the detector efficiency was also checked when the fill voxels contained air instead of water. In this case an increase of detector efficiency was observed. This leads to the assumption that mainly the effect of varying the mass between lung and detector plays a role when eroding the lungs.

The lung eroded by one voxel layer was taken as a reference for the following displacements. The displacements in x-direction (right-left) including the movements of the lobes towards and away from each other showed very little effect on the total detector efficiency. Detailed results for each single detector are not presented in this thesis, but losses in the phoswich detector 1 were compensated almost perfectly by detector 2, so that the sum of both detector efficiencies was almost constant.

The displacements in y-direction (posteroanterior) showed the most significant effect on the detector efficiency. These displacements were achieved through variations of the chest wall thickness (CWT) and variations of the distance from the center of gravity from the source to detector. Both effects (CWT and distance) together resulted in much higher variations compared to the effects observed by erosion. The displacement from +4 mm to –4 mm from the reference scenario resulted in +7.4% to –7.3% changes in the detector efficiency. Displacements in z-direction (superoinferior) of –4 mm to +4 mm showed fluctuations in the detector efficiency between –2.1% to +1.9%.

Conclusion

There is a problem as the lungs' center of gravity moves with each erosion, so the method cannot be used to examine the influence of erosion separately. Nevertheless, the experiments have shown that distance and CWT are the major parameters effecting detector efficiency for lung counting. While right-left-organ-displacements are compensated by the use of two detectors covering the a wider area above the chest, the superoinferior movements are quite sensitive and should not be neglected. Due to the methods, only small organ displacements of one voxel dimension could be performed. Further investigations were necessary.

6.1.3 Novel Method for Investigating Various Parameters

Section 6.1.2 described a voxel-wise displacement of the lung as the source organ. The center of gravity was not controllable and the displacements depended on the voxel dimensions. To overcome this, a second more comprehensive investigation was started to check the influence of various parameters in detail.

Besides the lung counting simulation additional liver counting was introduced and besides ^{241}Am, ^{137}Cs was used as a radiation source to investigate energy dependencies. Cesium usually does not accumulate in the lung or liver after uptake, but after a few days it is widely distributed throughout the whole body and slightly accumulated in muscle tissue as an alkaline metal with chemical properties like potassium. The LLNL Torso Phantom model (section 4.7.1) was used for this set of simulations, so the four overlays could also be used to vary CWT.

The results for all parameters investigated are discussed in the next sections. The following overview gives a summary of the parameters that were varied:

- organ (left and right lung lobes, complete lung, liver) with two detector positions (lung and liver),

- radionuclide (^{241}Am and ^{137}Cs),

- CWT (no overlay and 4 different overlays composed of 3 different materials, see section 6.1.4),

- organ mass (three different values generated due to shrinking, see section 6.1.6),

- organ displacement (in five steps in three directions, see section 6.1.5),

- detector displacement (in three steps in one directions, see section 6.1.5),

- fill material (to fill the empty space resulting from the organ shrinking, see section 6.2.1), and

- lung density, see section 6.2.7).

Method Details

The organ transformation was achieved through the functionalities provided by ITK[1] (Kitware, Inc., 2009c; Ibanez *et al.*, 2005). The target organ was separated and then transformed relatively to its center of mass by a 3×3 matrix plus a translation vector by ITK's affine transformation (see section 7.2.3, equations 7.1 and 7.2). The diagonal elements of this matrix contained the shrinking factor. Other matrix elements were all

---

[1]Insight Toolkit

set to zero. The translation vector describes the movement of the organ. To revoxelize the organ in the original grid fitting into the original voxel phantom, a resampling filter was applied.

In the original voxel model the target organ was removed by redefining its voxel to three different tissue substitutes as a filling material: the 87/13, the 50/50, and the 0/100 adipose/muscle-tissue equivalent as introduced for the LLNL Torso Phantom in table 4.3. Then, the transformed and revoxelized target organ was added to the original voxel model. At this step a check was performed to avoid overlapping with tissue other than the chosen filling tissue.

The liver was used in its original size and then shrunken with the factors 0.9328 and 0.8547. These factors were chosen to match particular mass values described in ICRP 23 and 89 (ICRP, 1975, 2002). The original liver had a mass of approximately 2200 g, which represents the 80th percentile value for male adults according to Boyd in ICRP 23. The second liver size has a mass of approximately 1800 g (ICRP reference male adult), and the third one 1400 g (ICRP reference female adult), respectively.

The lung lobes were used in their original size and then shrunken by the factors 0.92 and 0.8512, which resulted in complete lung masses of 1200 g for the original (ICRP reference male adult), 950 g (ICRP reference female adult), and 750 g (15-year old girl), respectively. Despite the shrinking, the center of mass could be kept within ± 0.3 mm for all coordinates in this investigation. A perfect constancy could not be reached due to the resampling filter.

The translational movements were set to nominal values of -8 mm, -4 mm, 0, $+4$ mm, and $+8$ mm in all three directions, but the actual translation did not reach the desired nominal values due to the resampling filter. The translations of the center of mass had slightly lower absolute amounts than 8 or 4 mm. This was fully considered in the analyses of the results.

Detector Position and Setup

The detector position for liver and lung counting was the following: the right-left-position (x-direction) of the two detectors were set to be symmetric from to the center of gravity of the whole voxel model excluding air voxels. The distance from each detector center from that point was 117 mm, while one detector was shifted to the positive x-direction, the other one was moved to the negative x-direction.

The detectors were moved in superoinferior-direction (z-direction) towards the center of gravity of the measured organ (lung or liver). The detectors were positioning in the standard position for lung and liver counting used at IVM, i.e. they were inclined by a rotation of 25° (−25°, respectively), to be parallel to the surface of the phantom. In posteroanterior-direction (y-direction) the detector were moved 1 mm above the phantom surface with the largest overlay (overlay 4). The detector was kept at this position for all measurements for the respective organ with one exception. To examine the effect of distance the detector was moved in y-direction to the center of gravity of the organs for the specific setup with the voxel model of the LLNL Torso Phantom.

The liver, the left lung lobe, the right lung lobe as well as both lung lobes together were used for this examination. All four overlays had been used. In total 680 simulations were performed with an approximated computing time of 2380 hours.

The number of simulated particle histories was 5×10^6, which resulted in relative statistical counting errors (equation 3.3) of less than 0.43% for each detector in the ROI. For ^{241}Am the ROI was 20 to 80 keV and for ^{137}Cs the ROI was 598 to 727 keV (± 1.25 FWHM around photo peak). The phoswichs' CsI(Tl)-crystals were used for counting ^{137}Cs-photons, since the energy goes beyond the adjusted energy range of the NaI(Tl)-crystal.

Analysis of Results

Enough results were produced to perform a sound linear regression of the results. Linear relationships could be assumed for the examined ranges and showed good correlations (squared Pearson Correlation Coefficient, i.e. R^2-value) for most parameter pairs. Often relative values are applied for better comparison of the absolute results. Microsoft Excel was used to plot the results and calculate the regressions. It was decided not to display error bars from statistical counting errors, because such errors were always clearly below 0.5% hence hardly visible in the plots.

A selection of meaningful results are presented in the following sections.

6.1.4 Chest Wall Thickness

For lung and liver counting the chest wall thickness (CWT) is of major importance (Dean, 1973; Fry and Sumerling, 1980). The CWT was varied in the experiments described in section 6.1.3. In summary, four different overlays were superimposed onto the plain voxel model to change the CWT. It should be noted that unlike for real measurements, the detectors were fixed. Hence, CWT was the only parameter that changed. The CWT values of the overlays came from the phantom's specifications (section 3.3.2), the densities came from table 4.6 and the material from table 4.3.

Results

Figure 6.5 shows the relationships of CWT and detector efficiency for the combinations lung and ^{137}Cs, as well as liver and ^{241}Am. The relationships can be described well with linear functions. A decrease of roughly 26% (21%) for ^{241}Am (^{137}Cs) for both liver- and lung detector counting efficiency for 23 mm additional tissue on the chest was observed.

Figure 6.6 shows the influence of posteroanterior-(y)-displacements and the different CWT in three dimensions. The relative changes of detector efficiency between the different y-displacements are similar for all overlays.

Table 6.3 shows the slopes determined by linear regression for the shrunken lung by factor 0.8512 for different organ shifts and overlay materials. The relative changes for detector efficiency seem to be almost constant for one overlay material.

Figure 6.7 illustrates the ratio of the energy bins of the efficiency spectrum without and with overlay 4. Not in all energy bins the ratio decreases to the same extend when an overlay is applied. For ^{137}Cs, the range from 500 to 600 keV has even a ratio higher than one. The main photo peaks are weakened by the overlay, but scattering at lower energies

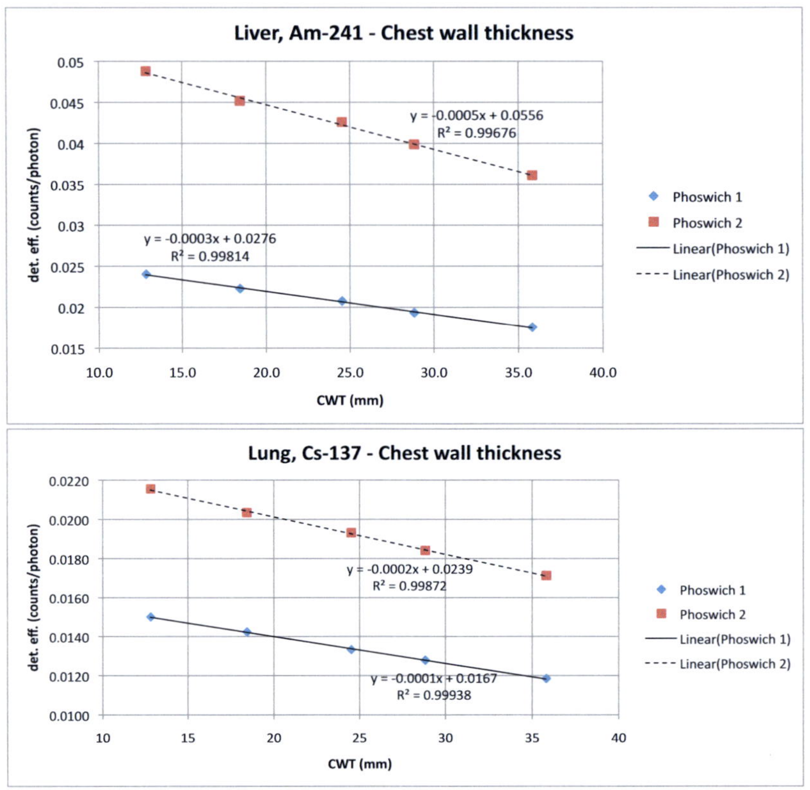

Figure 6.5: Detector efficiency plotted against CWT. The top plot shows the relationship for the original liver with ^{241}Am. Below, there is a plot of a ^{137}Cs scenario of the original lung.

is partly enhanced. The ROIs for ^{137}Cs and ^{241}Am are different from this point of view. The ^{137}Cs ROI does not include such scattering energy ranges, while the chosen ROI for ^{241}Am is broad enough to include the scattering range around 40 keV.

Conclusion

Depending on the composition of the chest wall tissue and the radionuclide, one can assume that roughly 10 mm additional CWT will result in a decrease of detector efficiency 10 to 13%. Adding CWT increases

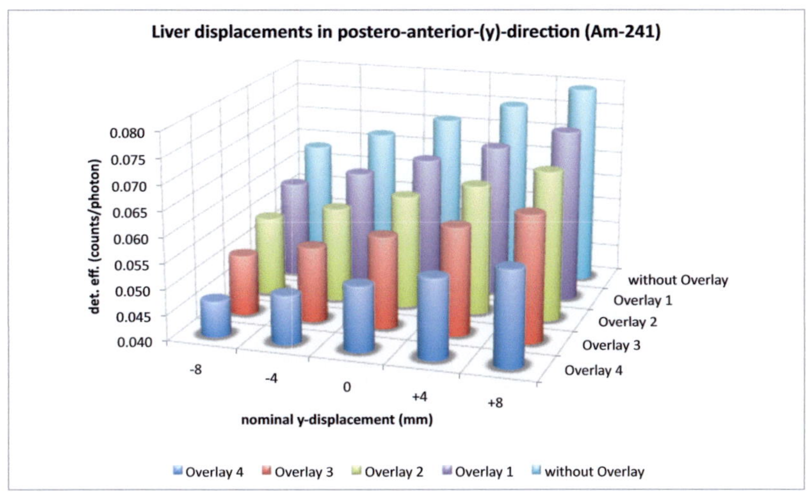

Figure 6.6: CWT and y-displacements for liver counting scenarios.

scattering in the energy ranges below the main photo peak.

At IVM an empirical formula (equation 3.4) is used to determine the CWT (see section 3.3.3). It is based on the correlation of the ratio of body mass and height with ultrasound measurements of the CWT. Ultrasound measurements are not perfectly reproducible. Deviations of up to 7 mm have been reported when different operators performed the measurement (Fry and Sumerling, 1980). Uncertainties of empirical formulas for the determination of CWT have been reported to be up to 5 mm (Dean, 1973). Every millimeter of CWT uncertainty results in at least 1% error for the detector efficiency.

Table 6.1: Detector efficiencies caused by translations and rotations of phoswich detector 1 in a lung counting scenario with ^{241}Am.

Variation of parameter			Detector eff. (counts/photon)	Relative to the reference
Reference			0.0206	1.000
Δx	=	−1.0 cm	0.0217	1.060
Δx	=	−0.5 cm	0.0212	1.025
Δx	=	+0.5 cm	0.0201	0.974
Δx	=	+1.0 cm	0.0196	0.948
Δy	=	−1.0 cm	0.0220	1.063
Δy	=	−0.5 cm	0.0213	1.032
Δy	=	+0.5 cm	0.0201	0.971
Δy	=	+1.0 cm	0.0194	0.941
Δz	=	−2.0 cm	0.0197	0.956
Δz	=	−1.0 cm	0.0203	0.981
Δz	=	+1.0 cm	0.0210	1.015
Δz	=	+2.0 cm	0.0211	1.020
ΔW_h	=	−10°	0.0208	1.008
ΔW_h	=	−5°	0.0205	0.994
ΔW_h	=	+5°	0.0207	1.000
ΔW_h	=	+10°	0.0204	0.986
ΔW_v	=	−10°	0.0203	0.983
ΔW_v	=	−5°	0.0207	1.005
ΔW_v	=	+5°	0.0205	0.995
ΔW_v	=	+10°	0.0205	0.991

Table 6.2: Organ erosion and displacement of the MEET Man in $(4\ mm)^3$ voxels. The eroded lung voxels have been replaced with water voxels. The lobes were approached and departed in x-direction by 1 voxel.

Lung modification (voxel)	Lung volume (L)	Sum det. eff. (counts/photon)	Relative to reference
original	3.75	0.04318	1.039
eroded by 1 (reference)	2.97	0.04156	1.000
eroded by 2	2.27	0.04004	0.963
$\Delta x = +1$	2.97	0.04155	1.000
$\Delta x = -1$	2.97	0.04147	0.998
lobes by 1 departed	2.97	0.04143	0.997
lobes by 1 approached	2.97	0.04161	1.001
$\Delta y = +1$	2.97	0.04466	1.074
$\Delta y = -1$	2.97	0.03854	0.927
$\Delta z = +1$	2.97	0.04067	0.979
$\Delta z = -1$	2.97	0.04233	1.019

Table 6.3: Various lung-CWT scenarios with ^{241}Am and their relative effect on the detector efficiency. Each linear regression showed R^2-values above 0.997.

Detector	Linear regression slope (counts/photon/mm)	Relative to reference (%/cm)
	lung shrunken by 0.8512, no shift overlay material 50/50 adipose/muscle	
Phosw. 1	-2.03×10^{-4}	-11.8
Phosw. 2	-3.07×10^{-4}	-11.4
Sum	-5.10×10^{-4}	-11.5
	lung shrunken by 0.8512, y-shift by nominal −8 mm overlay material 50/50 adipose/muscle	
Phosw. 1	-1.76×10^{-4}	-11.5
Phosw. 2	-2.75×10^{-4}	-11.5
Sum	-4.50×10^{-4}	-11.5
	lung shrunken by 0.8512, y-shift by nominal +8 mm overlay material 50/50 adipose/muscle	
Phosw. 1	-2.25×10^{-4}	-11.5
Phosw. 2	-3.47×10^{-4}	-11.4
Sum	-5.72×10^{-4}	-11.5
	lung shrunken by 0.8512, no shift overlay material 87/13 adipose/muscle	
Phosw. 1	-1.83×10^{-4}	-10.6
Phosw. 2	-2.80×10^{-4}	-10.4
Sum	-4.63×10^{-4}	-10.4
	lung shrunken by 0.8512, no shift overlay material 0/100 adipose/muscle	
Phosw. 1	-2.21×10^{-4}	-12.8
Phosw. 2	-3.38×10^{-4}	-12.5
Sum	-5.59×10^{-4}	-12.6

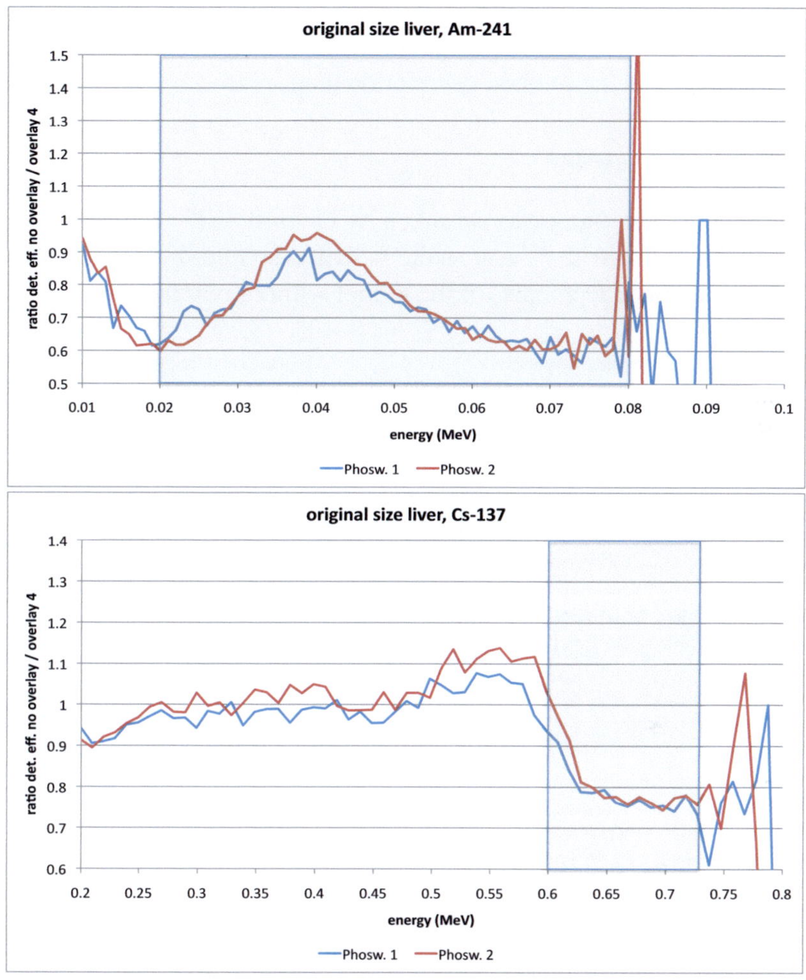

Figure 6.7: Two plots showing the ratio of the energy bins of the liver counting efficiency spectra without and with overlay 4 for the two investigated radionuclides for both phoswich detectors. The grey area marks the ROI for each radionuclide. The heavy fluctuations beyond the main photo peak are the results of poor statistics and can be ignored.

6.1.5 Continuous Displacements of the Source Organ

In this section the results for continuous displacements of organs are presented. This will give a more comprehensive insight than section 6.1.2 already gave. For the method description please refer to section 6.1.3.

Results and Discussion

First, the complete lung was shrunken by factor 0.8512 and then moved in three directions with ^{241}Am as the radiation source. Figure 6.8 shows results of the displacements for the model without overlays.

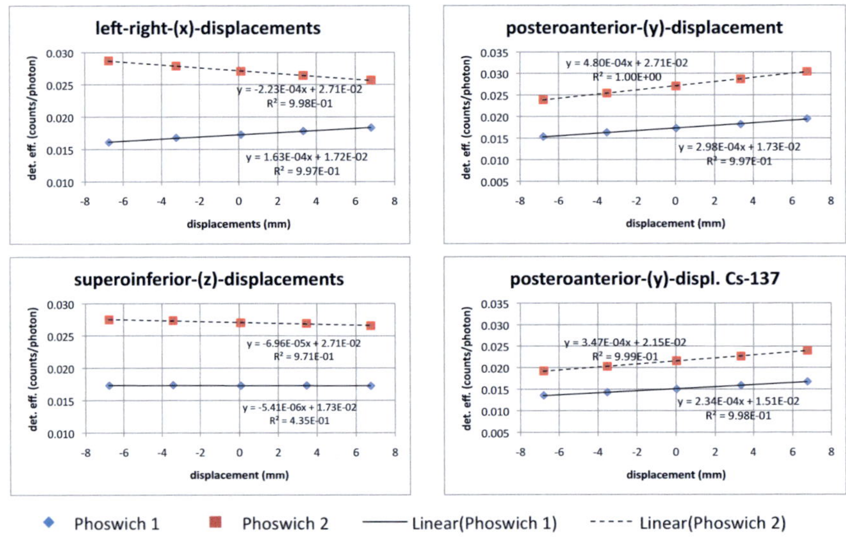

Figure 6.8: Results for continuous displacements of the complete lung (shrunken by factor 0.8512) and their influence on the detector efficiency for ^{241}Am. The right-bottom plot shows the behavior of the detector efficiency for ^{137}Cs with y-displacements.

The plots for the lung in figure 6.8 basically confirm the results presented in section 6.1.2. The shifts in x-direction for the single detectors cause considerable changes of detector efficiencies, while the sum of both detector efficiencies is almost constant. The total shift from −6.7 mm to +6.8 mm in y-direction caused a change on the detector efficiency from

−11.5% to +12.5%. The detector efficiency for ^{137}Cs for the same shifts changed from −10.4% to 11.1%. Generally the results for ^{137}Cs showed smaller deviations than those for ^{241}Am. Therefore it was decided not to present all ^{137}Cs results in case the relative variations are smaller than those from ^{241}Am.

The efficiencies for z-displacements from −6.8 to +6.7 mm result in efficiency variations from +1.2% to −1.1%. Good linear relationships were established and documented as R^2-values within the plots, except for the plot of detector 1 for the z-displacements, for which the values fluctuate with no distinct trend or slope.

The most sensitive parameter was the displacement in y-direction. Table 6.4 shows the slopes of the linear regression calculations performed to quantify the relative detector efficiency changes, when other parameters e.g. ratio of adipose and muscle tissue or overlay were varied. While an additional overlay has no effect on the relative variation of the detector efficiency, the fill material plays a role. The organ displacements in y-direction are basically a combination of two effects: changing the overlying tissue thickness and distance to the detector which can be comprehended easily by adding the amounts of the values from tables 6.3 and 6.5. It was necessary to repeat the detector displacement from section 6.1.1, as the voxel model was different and the effect needed to be related to the current reference scenario.

For the liver, the results are very similar. The respective results are summarized in figure 6.9. The absolute values are higher for the liver, but the pattern observed for the lung can be found here as well. Differences to the previous results from the complete lung are caused by the stronger asymmetry of the liver position. Therefore, there are stronger relative differences between the detector 1 and 2.

Results for the simulations with the separated left and right lung lobe bring no new knowledge. Therefore, it was decided not to discuss these results here.

Conclusion

The most sensitive parameter is the displacement in posteroanterior-(y)-direction. Considering a single detector, the right-left-(x)-direction is also

Table 6.4: Various scenarios of lung-y-displacements and their relative effect on detector efficiency (^{241}Am). The R^2-values of the linear regression were all greater than 0.998.

Detector	Linear regression slope (counts/photon/mm)	Relative to reference (%/cm)
no overlay, fill material 87/13 adipose/muscle		
Phosw. 1	2.98×10^{-4}	17.3
Phosw. 2	4.80×10^{-4}	17.8
Sum	7.74×10^{-4}	17.6
overlay 4. fill material 87/13 adipose/muscle		
Phosw. 1	2.15×10^{-4}	17.0
Phosw. 2	3.60×10^{-4}	18.0
Sum	5.75×10^{-4}	17.6
no overlay. fill material 0/100 adipose/muscle		
Phosw. 1	3.06×10^{-4}	18.2
Phosw. 2	5.00×10^{-4}	19.1
Sum	8.06×10^{-4}	18.8

quite sensitive, while the superoinferior-(z)-displacements are negligible, in contrast to the results in section 6.1.2. The geometry of two models (MEET Man vs. LLNL Torso Phantom) might be the reason for this disagreement. All observations are similar for both organs, the lung and the liver. The higher energy photons from ^{137}Cs cause a slightly less sensitive behavior.

Organ displacements consists of a combination of changing the thickness of the attenuating tissue (i.e. CWT in y-direction) and distance to the detector. Therefore the organ displacements in y-direction show the highest sensitivity observed in this investigation.

Table 6.5: Various scenarios of detector-y-displacements and their relative effect on lung counting detector efficiency. The R^2-values of the linear regression were all greater than 0.998.

Detector	Linear regression slope (counts/photon/mm)	Relative to reference (%/cm)
no overlay, no organ shift		
Phosw. 1	-1.16×10^{-4}	-6.7
Phosw. 2	-2.34×10^{-4}	-8.7
Sum	-3.50×10^{-4}	-7.9
overlay 1, no organ shift		
Phosw. 1	-1.03×10^{-4}	-6.3
Phosw. 2	-2.15×10^{-4}	-8.7
Sum	-3.18×10^{-4}	-7.8
overlay 2, no organ shift		
Phosw. 1	-9.40×10^{-5}	-6.4
Phosw. 2	-2.04×10^{-4}	-8.6
Sum	-2.98×10^{-4}	-7.8

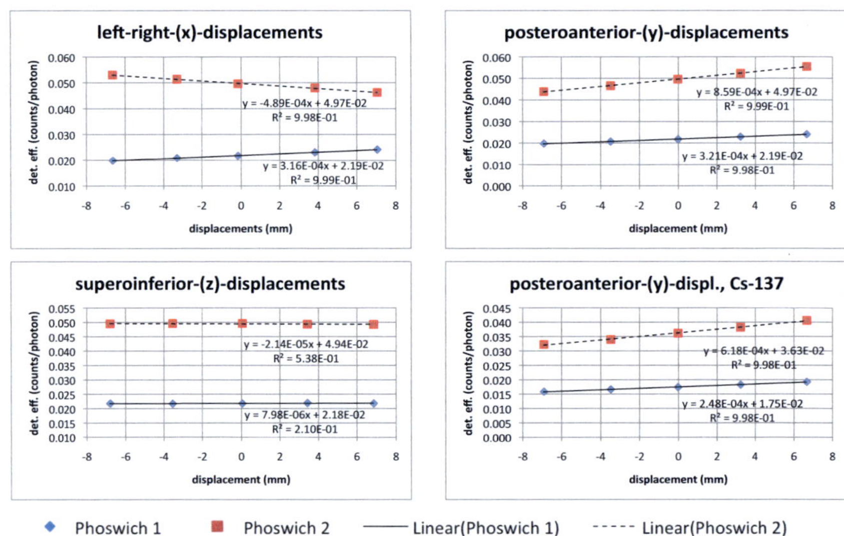

Figure 6.9: Results for continuous displacements of the liver (shrunken by factor 0.8547) and their influence on detector efficiency for ^{241}Am. The right-bottom plot shows the behavior of the detector efficiency for ^{137}Cs at y-displacements.

6.1.6 Organ Volume and Mass

The organ volume and mass were varied by shrinking factors in the simulations described in section 6.1.3. For the description of the method and the setup please refer to this previous section. The density of the organs was kept constant, so mass is proportional to volume.

Results

Exemplarily results for different lung masses are shown in figure 6.10. Additionally, the figure shows the data for different overlays. Linear regressions have been performed for all overlays. The lowest R^2-values for all overlays were 0.996 for lung (for both radionuclides) and 0.980 for liver using ^{137}Cs and 0.895 using ^{241}Am. A linear relationship of mass over detector efficiency can be assumed in the examined range.

For the loss of 465 g of lung mass a decrease of about 10% (11%) in detector efficiency was observed for ^{241}Am (^{137}Cs), suggesting that a clear energy dependency could not be observed. The liver lost 830 g (at about the same shrinking factor the lung lost 465 g), which resulted in a decrease of detector efficiency of approximately 2% (5%) for ^{241}Am (^{137}Cs).

As mentioned earlier (section 6.1.2), shrinking the lung lobes causes an increase of the mass between detector and organ, since the center of mass was kept at a constant position and the liberated voxels are filled with adipose/muscle tissue substitutes. The density of the liver is almost equal to the density of the filling material, while the lung's density is only approximately one third. This explains the stronger effect observed for the lung.

Following theoretical consideration helps to understand why the shrinking of the liver results in lower detector efficiency, even when the center of mass is kept constant. Figure 6.11 is a simplified sketch of the original sized organ and a shrunken organ at the same center of mass. Assumptions for the theoretical consideration are a one-dimensional problem with mono-energetic photons and no scattering besides the implicit attenuation by equation A.4. It can be proven that the transmission $T(x)$ of photons to the detector from the original organ is always greater than the transmission $T'(x)$ from the shrunken organ as described by equation 6.3. The transmission is assumed to be proportional to the detector efficiency.

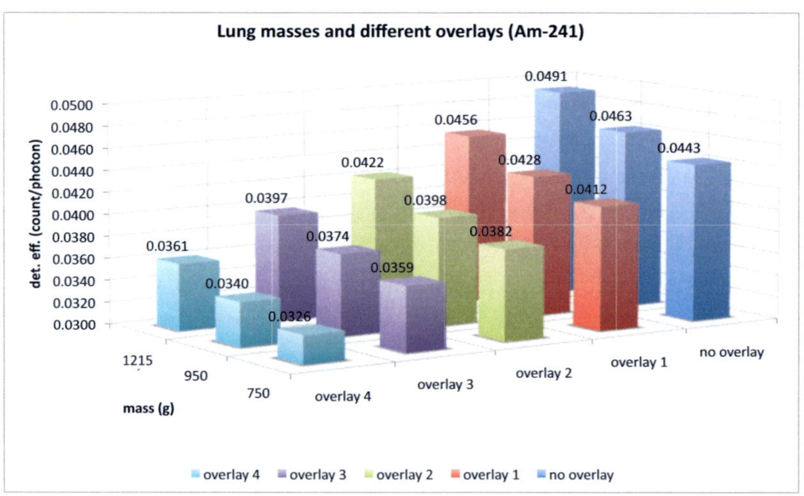

Figure 6.10: Sum of the detector efficiencies for the left and right phoswich for shrunken lungs (specified by their mass) in models with different overlays.

$$\text{with} \quad \frac{x_o'}{x_o} = s \quad \text{shrinking factor} \tag{6.1}$$

$$\frac{x_o - x_o'}{2} = x_t' - x_t = \Delta x_t \tag{6.2}$$

$$T(x) = \frac{I(x)}{I_0} = e^{-\mu x}$$

$$T(x) > T'(x) \tag{6.3}$$

$$T_a(x_a) \cdot T_t(x_t) \cdot T_o(x_o) > T_a(x_a) \cdot T_t(x_t') \cdot T_o(x_o')$$

$$T_o(x_o) \cdot T_t(x_t) > T_t(x_t') \cdot T_o(x_o')$$

$$T_o(x_o) > T_t(\Delta x_t) \cdot T_o(x_o')$$

$$\int_0^{x_o} e^{-\mu_o x} dx > e^{-\mu_t \Delta x_t} \int_0^{x_o'} e^{-\mu_o x} dx$$

$$\left(\frac{1}{\mu_o} - \frac{1}{\mu_o} e^{-\mu_o x_o} \right) > e^{-\mu_t \frac{x_o - x_o'}{2}} \left(\frac{1}{\mu_o} - \frac{1}{\mu_o} e^{-\mu_o x_o'} \right)$$

$$1 > e^{-\mu_t x_o \frac{1-s}{2}} \left(1 - e^{-\mu_o s x_o} \right) + e^{-\mu_o x_o}$$

$$1 > f(s, x_o, \mu_t, \mu_o) \tag{6.4}$$

$$\text{for} \quad x_o > 0, \quad s < 1, \quad \mu_t > 0, \quad \mu_o > 0$$

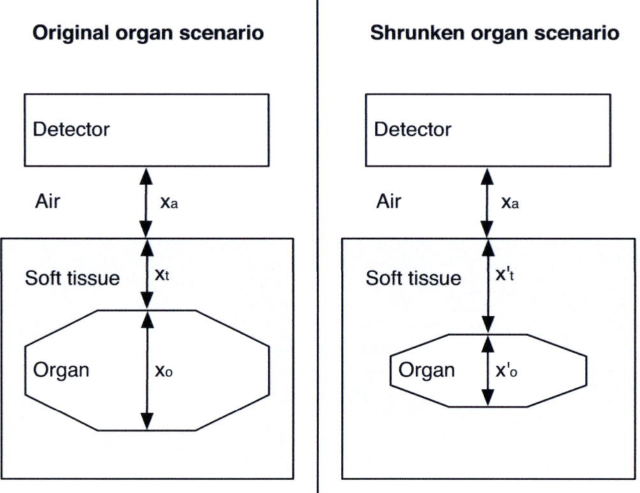

Figure 6.11: Simplified sketch of an original size and a shrunken organ. The center of mass of the organ is kept after shrinking by s (equation 6.1). Equation 6.2 determines the increase Δx_t of the soft tissue.

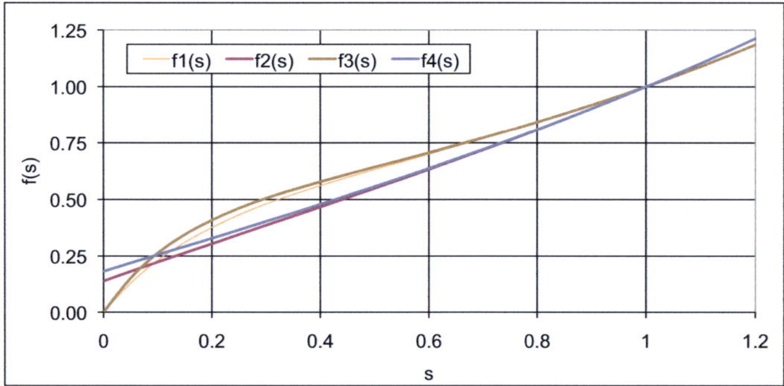

Figure 6.12: Plots of $f(s)$ (equation 6.4) for four different realistic parameter sets for lung and liver attenuation coefficients and photons at 60 and 100 keV. If s (equation 6.1) is smaller 1, one can assume (for the here presented simplified case) that $T(x) > T'(x)$ (equation 6.3), hence a shrunken source organ results in lower detector efficiencies.

Equation 6.4, plotted for realistic parameter sets in figure 6.12, presents the proof that a concentrated source distribution at the same center of mass results in a lower photon intensity and hence in a lower detector efficiency. Although such theoretical considerations can give a rough estimation about the examined situation, but they are only valid for simplified cases. For more complex and realistic calculations, one is forced to use Monte Carlo methods.

In addition, it is unusual that the results for ^{137}Cs at the lung and specifically for the liver showed stronger effects than for the low-energy emitting ^{241}Am. An explanation for this phenomenon can be found by looking at the definition of the ROIs. The ROI of ^{241}Am at IVM is usually defined from 20 to 80 keV to catch as many photons emitted from ^{241}Am as possible including contributions from scattering. The ROI for ^{137}Cs focusses on the main peak and not on the scattering energy range. The ROI is defined ± 1.25 FWHM around the main photo peak. Shrinking an organ in the body weakens the photo peak, but it promotes scattering.

Figure 6.13 shows the ratio of detector efficiency spectra from the original and the shrunken organs. It is illustrated that the ^{241}Am ROI covers the main peak and the scattering energy range. At 40 keV even more counts are observed, when the organ is shrunken. The scattering range of ^{137}Cs is not fully covered in the ROI, which means that the scattering effects do not play such an important role, when the organ is shrunken. This is the reason why ^{137}Cs is more sensitive at this point than ^{241}Am for the ROIs chosen in this examination.

Conclusion

For the liver the effect of organ mass on detector efficiency is small for the examined considerable changes in mass, or volume, respectively. The predominating effect is the decrease of detector efficiency caused by the more concentrated source distribution that can be lead back to the exponential attenuation law.

The tissue overlying the shrunken organs scatters the photons and the scattering is promoted by shrinking the organ. In case of the shrunken lung the liberated voxels were filled with material with a three-times higher density that causes the decrease of detector efficiency.

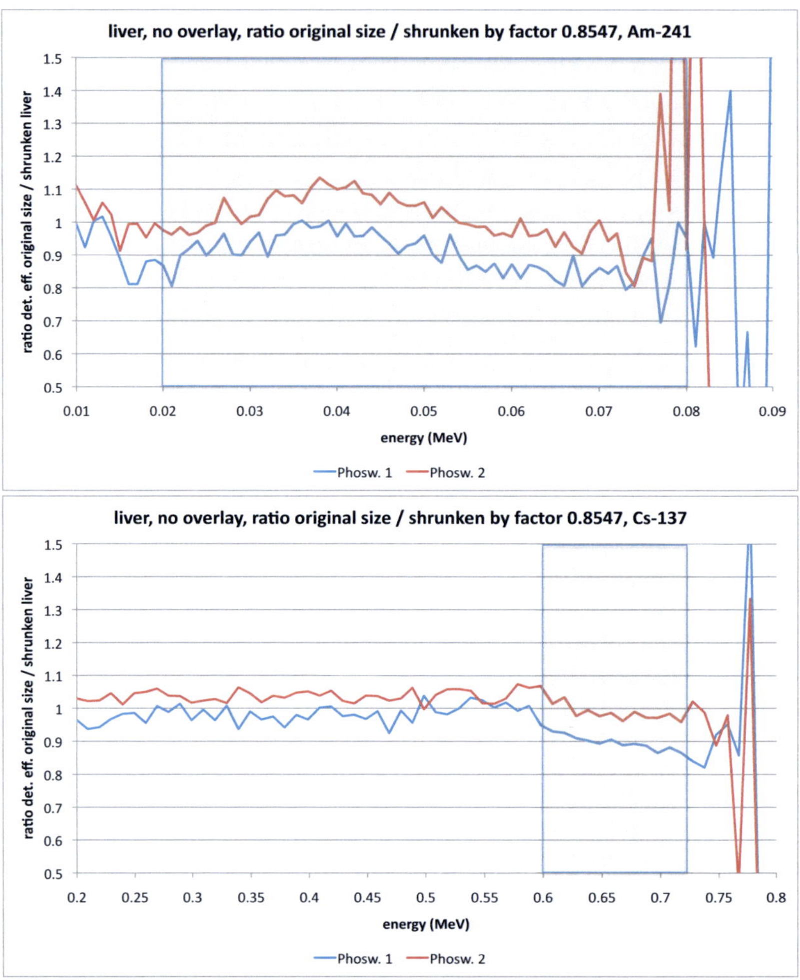

Figure 6.13: Two plots showing the ratio of the efficiency spectra of the original size liver and the lung shrunken by factor 0.8547 for the two investigated radionuclides for both phoswich detectors. The grey area marks the ROI for each radionuclide. The heavy fluctuations beyond the main photo peak are the results of poor statistics and can be ignored.

6.2   Special Parameters

6.2.1   Fill Material - Ratio Muscle / Adipose Tissue

Fill material was used to fill the liberated voxels resulting from shrinking the source organ in the method explained in section 6.1.3. Three different materials were used. The 87/13, the 50/50, and the 0/100 adipose/muscle-tissue equivalent. The difference of the three materials is basically the calcium content (see table 4.3). While the 87/13 adipose/muscle-tissue equivalent contains 0% calcium of its mass, the other tissue equivalent have 0.78% (for 50/50 adipose/muscle) and 1.7% for 100% muscle tissue equivalent.

Results

A linear trend (see figure 6.14) was observed when plotting the detector efficiency (^{241}Am) over the calcium content for the liver. The detector efficiency changes for the two extrema 0/100 and the 87/13 adipose/muscle tissue equivalent are approximately 2.7%. For ^{137}Cs scenarios, no trend was observed. The detector efficiency stays almost constant.

Tables 6.4 quantifies the effect of the fill material for lung scenarios for y-displacements of the organ. The detector efficiency changes for the two extrema 0/100 and the 87/13 adipose/muscle tissue equivalent are roughly 1.2%.

It was also shown, that there are little changes if the material for the overlays is changed. Table 6.3 shows results for different overlay materials. The changes observed in the ^{241}Am detector efficiency for the two extrema 0/100 and the 87/13 adipose/muscle tissue equivalent are approximately 2.2% for the sum of both phoswich detectors.

Conclusion

Effects from the different fill materials, resembling different adipose/muscle tissue ratios used in this investigation, are small for ^{241}Am and negligible for ^{137}Cs.

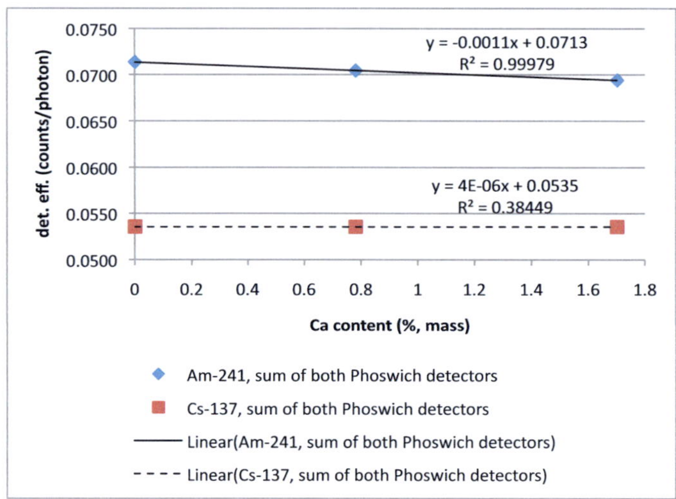

Figure 6.14: The plot shows the impact of calcium content in the fill material around the liver on the detector efficiency for ^{137}Cs and ^{241}Am.

6.2.2   Resolution and Voxel Size

The higher the resolution of a voxel model, the better it can reproduce anatomical details (see figure 6.15). A higher resolution means smaller voxels and a higher number of voxels. The memory and computing power limits of hard- and software are quickly reached when the voxel resolution increases. Often it is useful to use a low-resolution voxel model to save memory and time for simulations. Sometimes, anatomical details are required. An investigation with the MEET Man was performed to check the influence of model resolution on detector efficiency.

Method

The MEET Man model was used in 5 different resolutions in a lung counting scenario with ^{241}Am in the model of the PBC of the IVM. The phoswich detectors were placed onto the chest of the MEET Man models above the lung lobes in the standard configuration (25° inclined and rotated, parallel to the skin surface). The detector position was kept while the voxel model was exchanged with 5 different MEET Man models.

Figure 6.15: Sections from the MEET Man in two different resolutions. In coarse $(6\ mm)^3$ voxels (left image) and fine $(1\ mm)^3$ voxels (right image).

The cubical voxel sizes were $(2\ mm)^3$, $(3\ mm)^3$, $(4\ mm)^3$, $(5\ mm)^3$, and $(6\ mm)^3$ (see also table 4.1). The models have been cropped (head, abdomen, and part of the arms) in order reduce the number of voxels and to be able to load the model $(2\ mm)^3$ voxels completely into MCNPX. The torso, which is the relevant body part for lung counting, was kept. The highest resolution with $(1\ mm)^3$ voxels could not be used, as it was still too big for MCNPX after cropping.

The number of simulated particles was 10^7, which resulted in a relative statistical counting error of less than 0.22% in the ROI for ^{241}Am of 20 to 80 keV for each phoswich detector.

Results and Discussion

Table 6.6 summarizes the results. The changes are marginal, but a trend can be recognized. The finer the resolution, the lower the detector efficiency.

An explanation might be the algorithm used for coarsening the model. It is unknown what method the authors of the MEET Man used. Often an algorithm is used that is based on the mode, i.e. the most abundant voxel type in a cluster. Such algorithms cause an increase of the organ volumes which have a relatively small surface area. Volumes of other organ with a relatively large surface area, such as bones, blood vessels, etc., decrease. The MEET Man's bones show this effect. Bone attenuates low-energy photons very well due to its higher density and its high calcium content. Adipose tissue is a massive tissue with a relatively low surface area and a low density. It attenuates photons worse than other soft tissues in the body. Two columns in table 6.6 show the organ volume of bones and adipose tissue of the MEET Man models in different resolutions. It is suspected that the changes in bone and adipose volume contribute to the observed effect of decreasing the detector efficiency when the model resolution goes up.

Table 6.6: Detector efficiencies for lung counting with ^{241}Am calculated for MEET Man models in different resolutions. Additional columns list the organ volume of bone and adipose tissue.

Voxel edge length (mm)	Sum of detector eff. (phoswich 1+2)		Bone volume (L)	Adipose volume (L)
	(counts/photon)	relat. to 6-mm model		
6	0.04379	1.000	3.918	40.947
5	0.04327	0.988	4.112	39.933
4	0.04318	0.986	4.109	39.930
3	0.04251	0.971	4.126	40.196
2	0.04227	0.965	4.132	40.171
1	n/a	n/a	4.195	38.984

Conclusion

The voxel model resolution has an effect on the detector efficiency. The higher the resolution, the lower is the detector efficiency. There are indications that coarsening algorithms might have caused the decrease of bone volume with increasing voxel sizes and hence be responsible for the effect. The changes are small but their contribution as systematic errors can be reduced by choosing the highest voxel model resolution possible for hard- and software for simulations.

6.2.3   Scattering Effects of the Examination Table

It is assumed that the closest surrounding of the source plays a role on the detector efficiency due to scattering. A couple of simulations in the PBC have been performed with phoswich detectors to test this. The results illustrate the effect of scattering.

Setup

There were two setups:

a) An ^{241}Am-point source has been defined on the examination table of the PBC directly under phoswich detector 1. The distance to the detector was 30 cm. Detector 2 had the same y- and z-coordinate while the x-coordinate was shifted by -30 cm in x-direction. The detectors have not been rotated or inclined. The cylinder axes of the detectors were parallel to the y-axis.

b) The MEET Man in $(6\ mm)^3$ voxels was put onto the examination table of the PBC with the lung as ^{241}Am source. The two phoswich detectors have been placed over the model's lung lobes with about 1 mm distance to the chest's surface. The detectors have not been inclined or rotated, like for the point source.

The two setups a) and b) have been combined with three scenarios to test the point source and the voxel model for scattering effects. The first scenario was the most recent model of the PBC with the new, more detailed model of the examination table, i.e. artificial leather, foamed material, polyurethane plastic, and a steel frame. The second scenario was the *old* model examination table as defined by Doerfel *et al.* (2006), i.e. water with a density of 0.5 g/cm^3. The third scenario was setting all objects inside the chamber of the PBC to the material and density of air, i.e. the examination table and the shielding. Only scattering caused by air, by the voxel model (in case of b)), or by the detector models, is possible in this scenario.

The detector efficiencies were calculated in the ROI from 20 to 80 keV. The number of simulated particle histories was 10^7, resulting in statistical counting errors smaller than 0.4% for the six scenarios.

Results and Discussion

Table 6.7 shows the results for the counting efficiency of the two phoswich detectors. While the changes for the scenarios with the voxel model are marginal, the point source scenarios show considerable differences. The efficiency is 7.1% higher compared to the scenario with the old model of the examination table, and 26.2% higher compared to the scenario with the model and the shielding material of the PBC filled with air.

Besides the detector efficiencies also the spectrum shape changes. This is illustrated in figure 6.16. Scattering causes a shift of the peak center to lower energies and results in higher detector efficiencies.

Table 6.7: Scattering effects on phoswich detector efficiency caused by three different examination tables (ex. table) models in combination with a point source and a voxel model organ source.

| Scenario | Detector efficiency (counts/photon) | | | Sum relative to the |
	Phosw. 1	Phosw. 2	sum	new ex. table scenario
a) point source				
new ex. table	0.02035	0.00914	0.02949	1.000
old ex. table	0.01895	0.00846	0.02741	0.929
air	0.01513	0.00662	0.02175	0.738
b) voxel model				
new ex. table	0.02080	0.01937	0.04017	1.000
old ex. table	0.02081	0.01938	0.04019	1.000
air	0.02077	0.01934	0.04011	0.998

Conclusion

The model of the examination table and the shielding of the PBC plays a small role in case of simulating voxel models in lung counting position. For point sources on the examination table, this is however not negligible. The shape of the spectrum and the detector efficiency in the ROI changes considerably for point sources. The peak center is shifted towards lower energies. This must be considered particularly, when energy calibration is performed. The peak maximum might not coincide with the expected theoretical photo peak energy taken from literature.

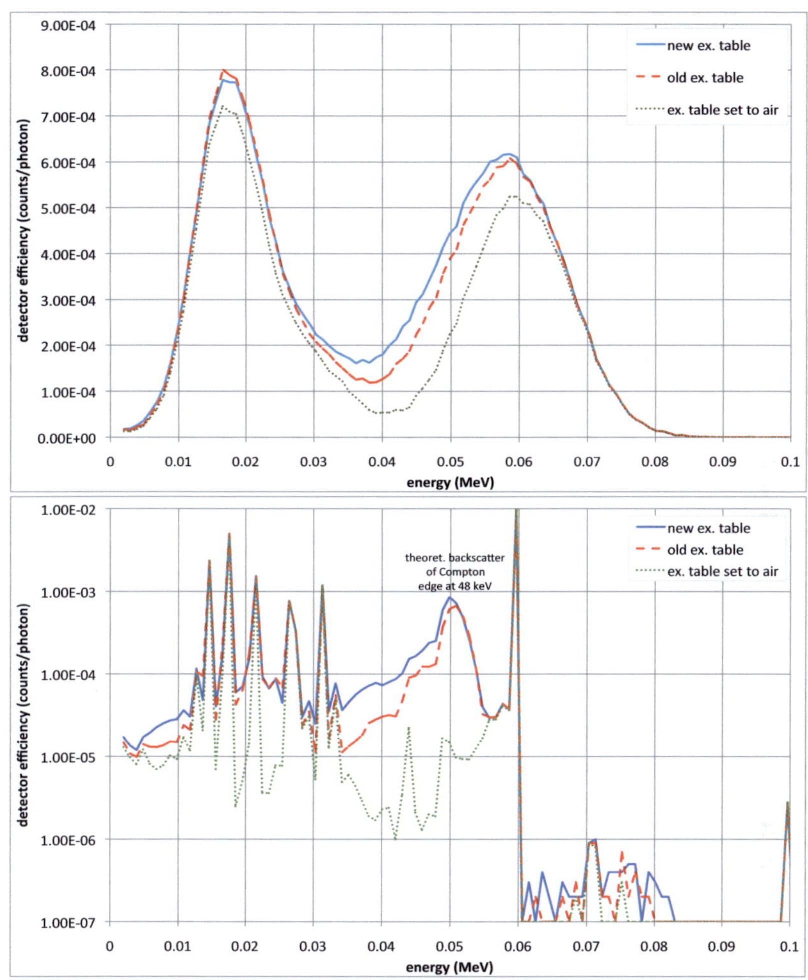

Figure 6.16: ^{241}Am-point source spectra of one phoswich detector with three different examination table models (old (Doerfel *et al.*, 2006), new (more detailed), and examination table filled with air). The upper plot shows the broadened spectra. The lower logarithmic plot shows the same spectra but without the broadening. Various X-ray peaks around 20 keV are now distinguishable from each other. It also shows the backscatter peak of the Compton edge at the theoretical energy (equation 5.3) of 48 keV for the old and new examination tables. The probability of Compton scattering in air is low, therefore this peak is missing the green-dotted spectrum.

6.2.4   Source Distribution

In this subsection the influence of the radionuclide distribution is discussed. The LLNL Torso Phantom is known to have a non-uniform density distribution in the lung lobes (figure 4.19). It was suspected that the density distribution was directly proportional to the activity distribution of radioactive lung lobes. To check the worst case scenario, two distributions have been investigated: a uniform radionuclide distribution in the lung volume and the radionuclides concentrated on the lung surface.

Setup

The LLNL Torso Phantom voxel model was used and positioned on the examination table of the PBC. Two phoswich detectors were placed onto the chest plate for lung counting about 1 mm above the surface with the typical standard inclination (25° inclined, 25° rotated) used at the IVM. The source was of choice ^{241}Am. In one experiment all lung voxel were assigned by MCNPX's `SDEF` card and in the other, only the lung voxel on the boundary lobes were assigned as source voxels. The algorithm used for defining the boundary source voxels was based on the the six neighboring voxels of a lung voxel in $\pm$x-, $\pm$y-, and $\pm$z-direction. If one of the six voxels was no lung voxel, it was defined as a source voxel. The number of simulated particles was 10^7, resulting in a statistical counting errors of less than 0.3%. Two different ROIs were defined from 20 to 80 keV and from 35 to 80 keV.

Table 6.8: Detector efficiency at different source distributions: the lung as a volume source and the lung as a surface photon emitter.

NaI(Tl) detector	ROI (keV)	Lung volume source (counts/photon)	Lung surface source (counts/photon)	Ratio volume/surface
left	20-80	0.02386	0.02679	1.123
left	35-80	0.02049	0.02288	1.117
right	20-80	0.03824	0.03703	0.968
right	35-80	0.03276	0.03160	0.965
left+right	20-80	0.06210	0.06382	1.028
left+right	35-80	0.05325	0.05448	1.023

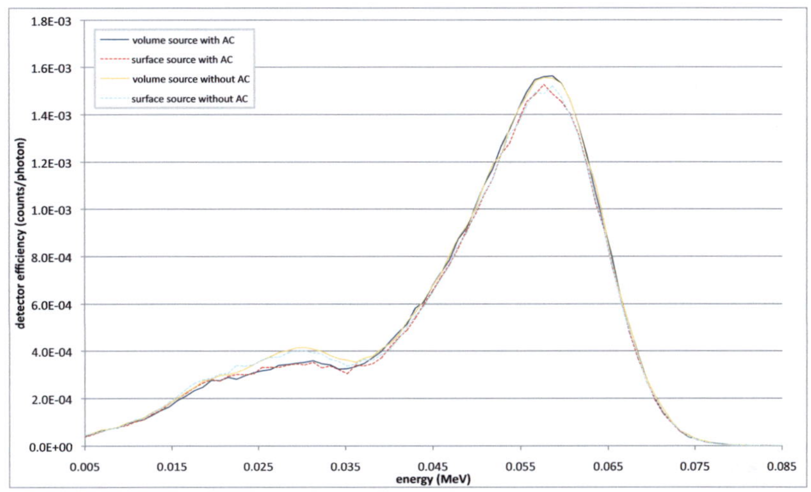

Figure 6.17: Simulated lung counting spectra of right Na(Tl) detector (phoswich 1) with lung as volume (solid lines) and surface source (dashed line) and with anti-coincidence feature and without. Only marginal differences were observed between the two different source distributions at the main peak. The anti-coincidence has a small effect in the energy range 10 to 40 keV. Tables 6.8 and 6.9 reveal the detector efficiencies.

Results and Discussion

Figure 6.17 shows spectra from each NaI(Tl) detector 1 of the phoswich system. When just the surface voxel were defined as source voxels, the ROI (20 to 80 keV) showed that the detector 1 over the left chest side had roughly 12% more counts, while the detector 2 over the right phantom side had about 3% less counts. The difference of the left and right detector can be explained by the asymmetry of the two lung lobes. The left lung lobe is obscured by the heart, hence it there is a lower counting efficiency on the detector above. The increase of 12% is possibly caused by a greater surface area of that lobe in the front part of the chest due to the distinct *wrapping* of the lung lobes around the heart. The summed ROI of both detectors resulted in a change of about 3% more counts compared to the scenario with the volume lung source.

Conclusion

Detector 1's efficiency increased considerably whereas the opposite was observed for the other detector. Obviously, it is an advantage using two detectors, so that such effects can be averaged by calculating incorporated activities with the sum of both detector efficiencies.

The worst-case scenario investigated here, is surely an exaggeration. The source distribution in the active lung lobes of the LLNL Torso phantom is more uniformly distributed than concentrated on the surface. The true detector efficiency lies somewhere between the determined values.

6.2.5 Anti-Coincidence

The same setup as described in section 6.2.4 was used to show the influence of the anti-coincidence (AC) in different ROIs simulated by MCNPX 2.6. The MCNPX version 2.5 has a documented bug about the energy cutoff for Auger electrons when using the specific F6 tallies (Hendricks *et al.*, 2008) and hence cannot calculate valid results when using a PHL card for the AC feature (McKinney, 2008). The new version of MCNPX 2.6 was not available at the ISF until January of 2009 at ISF, which explains why the AC feature could not be used for most experiments performed in this work. The investigation of the AC feature was however necessary to check if the conclusions from the results calculated without AC are still valid.

Results

Figure 6.17 illustrates the influence of the AC feature. Table 6.9 shows the results using the same setup as for the results displayed in table 6.8. Table 6.10 shows the ratios of the results.

As expected the tally bins without AC feature had more counts, roughly around 2% in the ROI from 20 to 80 keV. In the second ROI ranging from 35 keV to 80 keV, the difference is less than 0.5%. The main effect takes place between 10 and 40 keV. The photo peak is just slightly affected on the lower energy side. The corresponding values observed for the two detectors are similar. The ratios in table 6.10 are

Table 6.9: Detector efficiency for different source distributions: the lung as a volume source and the lung as a surface photon emitter with AC. Same setups as for table 6.8

NaI(Tl) detector	ROI (keV)	Lung volume source (counts/NPS)	Lung surface source (counts/NPS)	Ratio volume/surface
left	20-80	0.02340	0.02629	1.123
left	35-80	0.02040	0.02277	1.116
right	20-80	0.03750	0.03632	0.969
right	35-80	0.03266	0.03150	0.964
left+right	20-80	0.06090	0.06261	1.028
left+right	35-80	0.05306	0.05426	1.023

Table 6.10: Ratios of detector efficiencies with and without AC, calculated from the data of tables 6.8 and 6.9.

NaI(Tl) detector	ROI (keV)	Ratio (volume source) without/with AC	Ratio (surface source) without/with AC
left	20-80	1.020	1.019
left	35-80	1.005	1.005
right	20-80	1.020	1.019
right	35-80	1.003	1.003
left+right	20-80	1.020	1.019
left+right	35-80	1.004	1.004

almost constant within the chosen ROIs. It seems that they do no depend either on the detector (left or right) or the source distribution (volume or surface). It can thus be assumed that these ratios are also constant for other scenarios.

Discussion

In low energy ranges below 100 keV, the Compton scattering does not interfere the photo peak area (see equation 5.3). Other simulations with high-energy emitting radionuclides, which are not described here, showed that the AC feature has also a minor effect on the detector efficiency (see figure 6.18). As long, as the ROI just covers the main photo peak, a negligible influence of the AC is observed. Care must be taken, when

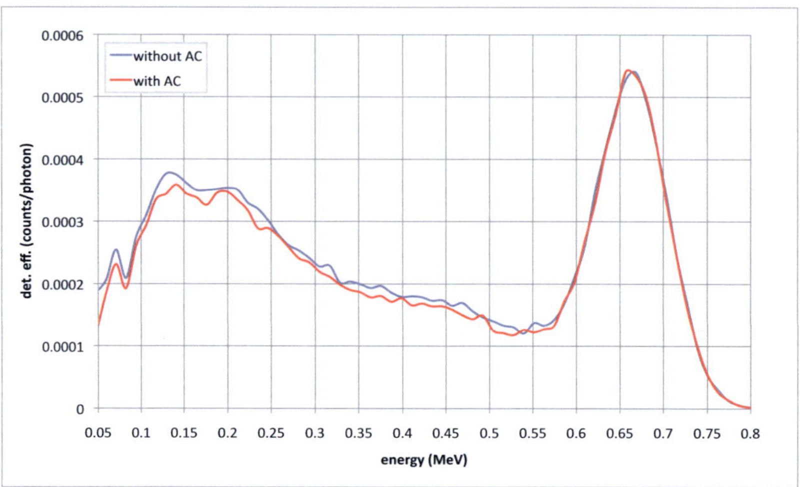

Figure 6.18: Example comparison of one spectrum simulated with and one without AC feature. The spectra are the result of a lung counting scenario involving a ^{137}Cs source in the lung of a voxel model. While the main photo peak is almost unchanged, the continuum below is reduced when the AC feature is turned on.

the main photo peak of interest sits on top of a scattering continuum of another peak of higher energy. The AC feature reduces this continuum, hence the gross counts in the peak area of interest.

Conclusion

All relative conclusions drawn from investigations without AC feature are valid for those with AC. Relative results are not affected, only absolute values will be shifted to lower detector efficiencies when using the AC feature. The extend of this shift depends on the ROI. If the ROI covers just the main photo peak, minor effects are expected. Generally, possible scattering continua must be considered and subtracted to obtain the net peak area of interest anyway with or without using the AC feature. In this thesis, secondary photo peaks were not considered, where the AC feature could have a more severe influence such as the 30 keV peak of ^{137}Cs.

6.2.6 Segmentation Errors

A problem observed during segmentation of the LLNL Torso Phantom CT data was the threshold setting of the lung tissue substitute in order to distinguish between lung tissue and air (see section 4.7.3). Setting the limit on the Hounsfield scale to high or too low had consequences on the volume of the lung lobe segments. The mass was weighed with a balance, hence the density was adapted from the mass over volume ratio. The following examination gives an impression on how the lung counting detector efficiency for ^{241}Am changes with inaccurate segmentation and hence inaccurate volume of the lung lobes.

Method

The model of the LLNL Torso Phantom was used in MCNPX simulations. The boundary voxel layer of the lung lobes was eroded and replaced by air voxels. Air is the main neighboring voxel type of the lung lobes for the LLNL Torso Phantom.

A typical lung counting simulation scenario was set up. Phoswich detectors 1 and 2 of the PBC were placed about 1 mm above the chest surface of the model in the standard position (25° rotated and 25° inclined). ^{241}Am was chosen to be the source radionuclide. The ROI was 20 to 80 keV. The number of simulated particles was 10^7, resulting in statistical counting errors of less than 0.4% for a single detector.

The original lung volume was 3702 mL. It was reduced by 12.8% (24.7%) by eroding one (two) voxel layer(s) from the organ boundary. Once the density was adapted to the new volume, and in a different setup, the reference density of the original lung was taken. The center of mass was only calculated, since the erosion of voxel layers of not point-symmetric organ results in a shift of its center of mass.

Results and Discussion

Table 6.11 summarizes the results. It is noticeable that the center of mass moves in negative x-, y-, and z-direction with each erosion step. But the moves a marginal, only small contributions on detector efficiency are expected. The most sensitive direction is the y-direction. According to this, we would expect a decrease in efficiency, which is not the case when the density is kept constant. The enlargement of the air gap between chest cover and lung is causing an increase of detector efficiency. Adapting the density at the same time over-compensates this effect. The deviations are increasing when two voxel layers are eroded.

Table 6.11: Detector efficiencies of eroded lung lobes with adjusted and non-adjusted densities.

Voxel model	Volume difference vs. reference (%)	Density (g/cm^3)	Center of gravity (x, y, z) (mm)	Efficiency, absolute (cnts/phot.)	Efficiency difference vs. reference (%)
Reference	0.0	0.326	228.1 120.7 205.4	0.05190	0.0
single eroded lung volume, adjusted density	−12.8	0.374	227.5 120.3 204.3	0.05146	−0.8
single eroded lung volume, original density	−12.8	0.326	227.5 120.3 204.3	0.05238	0.9
double eroded lung, adjusted density	−24.7	0.434	226.7 119.8 203.3	0.05087	−1.9
double eroded lung, original density	−24.7	0.326	226.7 119.8 203.3	0.05275	1.6

Conclusion

A segmentation error resulting in a volume deviation of more than 10% is unrealistic, since an unambiguous histogram was available (see figure 4.14). The adaptation of the lung density over-compensated possible errors. This investigation does not clarify, if the shifts of the center of mass caused the marginal changes in detector efficiency, or if it was caused by the density and volume changes. Nevertheless, the deviation of detector efficiency for the single eroded lung are in the marginal range of less than 1%, hence small segmentation errors are negligible.

6.2.7 Breathing - Density Changes

The process of breathing combines many variable parameters that have an effect on detector efficiency.

- changes of the density of the lung due to breathed air,

- changes of the volume of the lung,

- changes of the distance of the lung's center of mass to the detector, and

- deformation of the chest.

The process of breathing has a considerable effect on the density of the lung lobes. The mass of the breathed air is marginal, but the volume can be several liters. An average lung density is hard to define and to determine, therefore it is a source of uncertainty. Changes in the organ volume and their effect on detector efficiency have been discussed in section 6.1.6 but the breathing process cannot be compared with that. With breathing the whole chest deforms, but the CWT stays almost constant, as the chest tissue is not compressible. Hence the chest is deformed. The center of mass moves with breathing motion. Researchers at RPI (Zhang *et al.*, 2006) developed a technical model of the breathing motion of the chest. This examination was focussed on technical possibilities rather than anatomical realism. Further investigations are necessary to determine the motion and the geometry of the chest in different breathing phases.

Nevertheless, one of the breathing parameters, namely the change in density, can be examined easily. In the following Monte Carlo analysis, the density of the lung was varied. In contrary to real breathing, the chest of the model was not deformed and the volume of the lung was not changed.

Method

The setup and method are already described in section 6.1.3. Besides the original density of the LLNL Torso Phantom, the lung density was varied from the 0.25 g/cm^3 (lung density from reference male adult (ICRP, 2002)) up to 1.05 g/cm^3 (airless lung parenchyma density (ICRP, 2002)).

Results and Discussion

Figure 6.19 and table 6.12 illustrates the result. A higher density results in lower detector efficiency due to self-attenuation of the radiation in the lung tissue. The relationship can be approximated to be linear.

Table 6.12: Detector efficiencies for lung counting ^{241}Am with the LLNL-Torso phantom voxel model with different lung densities.

Density	Volume	Mass	Lung counting detector efficiency (counts/photon)			
(g/cm^3)	(mL)	(g)	Phosw. 1	Phosw. 2	Sum	sum rel. to ref.
0.327 (ref.)	3716	1215	0.01933	0.02975	0.04909	1.00
0.250	3716	929	0.01995	0.03066	0.05061	1.03
0.500	3716	1858	0.01809	0.02788	0.04597	0.94
1.050	3716	3902	0.01465	0.02272	0.03738	0.76

The tidal lung volume, i.e. the volume fluctuations caused by breathing, for an awake adult male in sitting position is approximately 0.75 L and for a resting, sleeping male adult, it is only 0.63 L (ICRP, 2002). The subject during a measurement is lying but usually not sleeping, so the tidal volume is assumed to be in the range from 0.63 to 0.75 L. Applied to the LLNL Torso Phantom, this is roughly equivalent to a $\pm 10\%$ change in total lung volume and hence a density fluctuation of $\pm 10\%$, i.e. a density range from 0.294 to 0.360 g/cm^3.

The sum of the detector efficiency of both detectors for $\rho = 0.25$ g/cm^3 is about 3% higher than the original density of of the model, while for $\rho = 0.5$ g/cm^3 is was 6% lower. For the density range from 0.294 to 0.360 g/cm^3 the detector efficiencies can be derived from the functions given in figure 6.19. The detector efficiency changes about $\pm 1\%$ in this density range.

Conclusion

Breathing is a complex process and difficult to model. More detailed investigations are necessary. One part of the this complex process is the

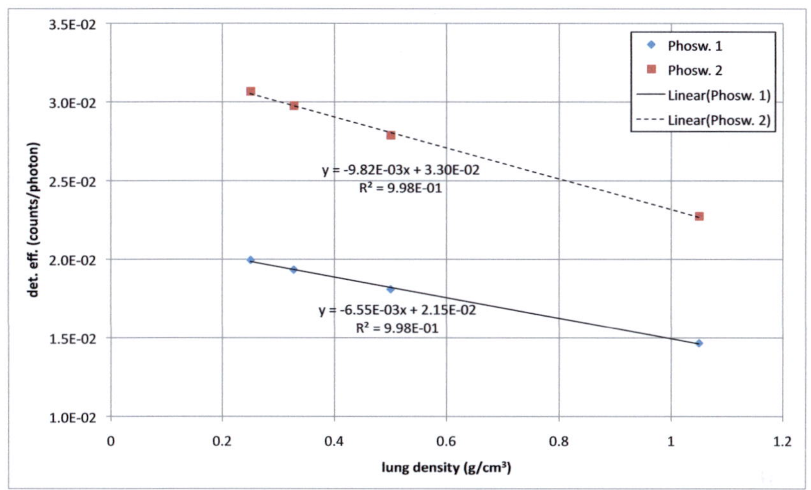

Figure 6.19: Lung counting detector efficiency for phoswich detectors plotted over lung density.

change of lung density. For an awake sitting male adult, the expected influence on lung detector efficiency caused by density fluctuations are marginal.

6.2.8 Breast Size

In this section the influence of different breast sizes of women on lung counting detector efficiency has been analyzed. The research presented in this section has been published by the author and others before (Hegenbart *et al.*, 2008).

Currently there are no physical female phantoms with adjustable breasts available, because these models are difficult to design and manufacture. Unlike male breasts, female breasts are difficult to standardize due to the large variety of possible breast sizes and shapes. Chest wall thickness (CWT) was discussed in section 6.1.4. It is known to have considerable influence on detector efficiency and appears to be linked to female breast size. Therefore a high sensitivity of detector efficiency was expected for this parameter.

Like most anthropometric data, breast dimensions depend on factors such as race, culture, age, gender, diet, and social status. An empirical formula was introduced by Berger and Lane (1985) to determine female CWT involving simple measurements of the chest circumferences in vertical and supine positions. However the parameters in the formula by Berger and Lane (1985) may need to be reviewed as increases in body weight over the last few decades have caused considerable changes in breast sizes of women (Lee *et al.*, 2007c; Odgen *et al.*, 2006).

Method

This work uses the deformable RPI Adult Female model (Xu *et al.*, 2008) based on BREP surface meshes (section 4.5.3). The model was deformed to create a series of models with different breast sizes.

The breast size was quantified according to European standards (CEN, 2001, 2002, 2004). The quantifying dimensions are bust girth (BG), which is the maximum horizontal girth, the underbust girth (UBG), and the resulting secondary dimension called cup size (CS) that is the difference between BG and UBG according to equation 6.5. For bras, the range of the CS is usually coded with letters, e.g. *AA* for the range 10 to 12 cm, *A* for the range 12 to 14 cm, *B* for 14 to 16 cm, and so on.

$$CS = BG - UBG \tag{6.5}$$

A series of models have been generated with cup sizes ranging from AA to G. The deformation of the models was achieved by translation of the mesh vertices of the breast in their normal vector direction. Furthermore, adjustments were made to the breast shape in order to ensure a realistic result. The breast tissues were morphed with a 60° downward angle to simulate a gravity-caused tissue deformation. The models have been designed to represent a standing, bra-wearing woman. Additionally, a smooth transition between relocated and non-relocated vertices was achieved by softening the mesh with a Laplacian smoothing algorithm (Vollmer *et al.*, 1999). Bust girth and underbust girth were measured according to the definition (CEN, 2001) at the outermost breast skin contour using the rolling ball method (Armato *et al.*, 1999).

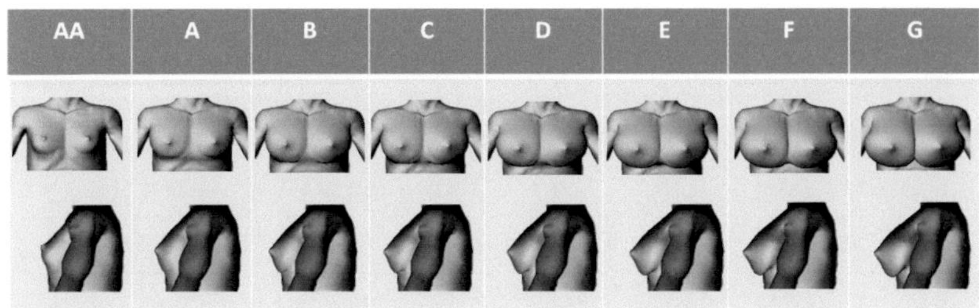

Figure 6.20: The chest of the different BREP models (standing, naked) viewed from the front and the left side (Hegenbart *et al.*, 2008).

The mass of the glandular breast tissue was kept constant for simplicity, but the tissue was aligned to the new nipple position. The only exception to this was the E-model. The glandular tissue was adapted to result in a glandularity of the breast of 40%. The letter code E40 refers to that model. The results for the glandularity are discussed in the section 6.2.9. The space between the glandular tissue and skin was filled with adipose tissue during the voxelization process (see next paragraph) of the organ meshes. The mass and center-of-mass of the adipose breast tissue and skin tissue changed because they are affected by the breast deformation. All other organ masses and their center-of-masses were kept constant. Table 6.13 lists relevant anthropometric data from the used models.

The series of deformed mesh models were then voxelized so that they could be adopted into MCNPX. The voxelization method is based on the parity count method together with the method of ray stabbing on a polygon surface (Nooruddin and Turk, 2003). The voxel dimension of the models was chosen to be cubical with an edge length of 3 mm, resulting in a total number of voxels of approximately 22×10^6.

The density and elemental composition of all the voxelized tissues and organs, except for the skin, were defined according to ICRP (2002). For the AA-model, the density of the skin was reduced from its nominal value of 1.09 g cm^{-3} in order to comply with the mass of the skin recommended by ICRP (2002). Without this adaption, the skin mass would be too high, because the small dimensions of the skin tissue cannot be described accurately with cubic voxels of side 3 mm.

Table 6.13: Relevant anthropometric data of the models sorted by cup size.

CS letter code / model	BG (cm)	UBG (cm)	CS (cm)	Glandularity (%)	Mass of breasts (g)
AA	99.6	89.2	10.4	40	500
A	102.6	89.2	13.4	15	1317
B	104.6	89.2	15.4	12	1700
C	105.8	89.2	16.6	11	1897
D	107.6	89.2	18.4	9	2331
E	110.0	89.2	20.8	7	2791
E40	110.0	89.2	20.8	40	2854
F	112.0	89.2	22.8	6	3304
G	114.0	89.2	24.8	5	3855

Figure 6.21 illustrates the detector position relative to the model. The two phoswich detectors were positioned in superoinferior and left-right direction over the center-of-mass of the complete lungs (red axis in figure 6.21). Both detector axes are parallel to the anteroposterior direction (see also figure 6.4). The distance of the two axes in left-right direction was 210 mm. The detectors were not inclined or tilted. They were placed approximately 1 mm above the breast of each model.

Three radionuclides (^{241}Am, ^{137}Cs, and ^{60}Co) were chosen for the simulations to investigate possible energy dependencies. The ROIs for ^{137}Cs and ^{60}Co are chosen to be ± 1.25 FWHM around the peaks of interest. For ^{241}Am, the standard ROI 20 to 80 keV used at IVM was selected. A total of 5×10^6 particle histories were run for each simulation to achieve results with relative statistical counting errors less than 0.38%.

Results and Discussion

Table 6.14 shows the lung counting detector efficiencies for the radionuclide ^{241}Am.

It was renounced to list detailed results for ^{137}Cs and ^{60}Co. For higher photon energies, a slightly smaller sensitivity was observed, as illustrated in figure 6.22 showing detector efficiency in relation to cup

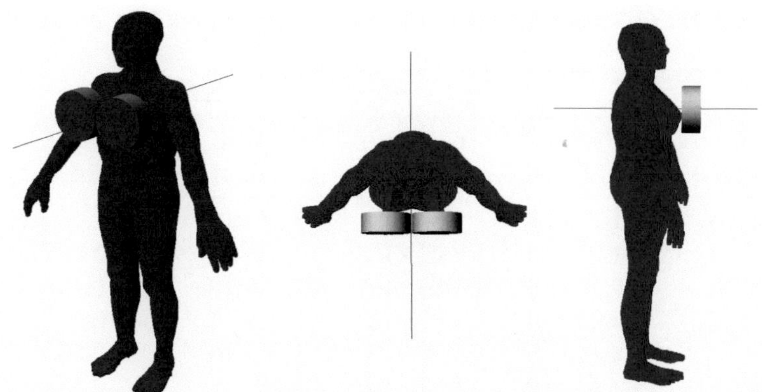

Figure 6.21: Phoswich detector position illustrated for the voxelized RPI Adult Female phantom with cup size C viewed from three different perspectives. A thin axis goes from the middle of the two detectors through the center of mass of the lung (Hegenbart *et al.*, 2008).

size. The relationships for each radionuclide can be approximated well with a power fit with R^2-values of at least 0.994%. Similarly figure 6.23 shows detector efficiency in relation to breast mass. The relationships for each radionuclide can be approximated well with a second order polynomial fit with R^2-values of at least 0.997%. For ^{137}Cs and ^{60}Co, the efficiency of the E40-model also fits well on each of the six fitted curves.

Conclusion

Especially for the low-photon emitter ^{241}Am, detector efficiencies decrease considerably with breast size. In the worst case, this can result in a reduction of detector efficiency of about 50%, when measuring ^{241}Am incorporation in a female lung, for a woman measured in a standing position with cup size G compared to a woman with cup size AA. If these factors are not considered, the calculation of the activity, and hence the estimate of the received dose, will be underestimated. Second order polynomial equations can describe the relationship of detector efficiency and cup size well. Power equations can describe the relationship of detector efficiency and total breast mass well.

Table 6.14: Calculated detector efficiencies for ^{241}Am in ROI 20-80 keV.

Cup size	Detector efficiency (counts/photon)			
letter code / model	phoswich 1 (left)	phoswich 2 (right)	total (sum)	sum relative to AA
AA	1.28×10^{-2}	1.50×10^{-2}	2.78×10^{-2}	1.00
A	1.04×10^{-2}	1.22×10^{-2}	2.26×10^{-2}	0.81
B	9.39×10^{-3}	1.09×10^{-2}	2.03×10^{-2}	0.73
C	8.93×10^{-3}	1.03×10^{-2}	1.92×10^{-2}	0.69
D	8.19×10^{-3}	9.40×10^{-3}	1.76×10^{-2}	0.63
E	7.49×10^{-3}	8.62×10^{-3}	1.61×10^{-2}	0.58
E40	7.29×10^{-3}	8.36×10^{-3}	1.57×10^{-2}	0.56
F	6.92×10^{-3}	7.84×10^{-3}	1.48×10^{-2}	0.53
G	6.53×10^{-3}	7.36×10^{-3}	1.39×10^{-2}	0.50

6.2.9 Breast Glandularity

The mammary glands, which are responsible for producing breast milk, are composed of glandular tissue. The term glandularity stands for the mass fraction of glandular tissue of the total breast mass. This is an important parameter for mammography dose calculations. Glandular tissue has a different elemental composition and a different density than the surrounding adipose breast tissue (ICRP, 2002).

In order to study the effect of glandularity on lung counting detector efficiency, the same setup and method as in section 6.2.8 was used.

The original E-model with a glandularity of 7%, was modified to generate a second E-model with 40% glandularity. Therefore this model is denoted as the E40-model. A value of 40% glandular tissue is recommended by (ICRP, 2002) for the reference female adult. Also Jamal *et al.* (2004) review data on glandularity and compared these with their own measured data. Observed glandularities ranged from 0.2% to 99.9% with a median of 46.6%. Younger women tend to have higher glandularity while obese women tend to have lower glandularity.

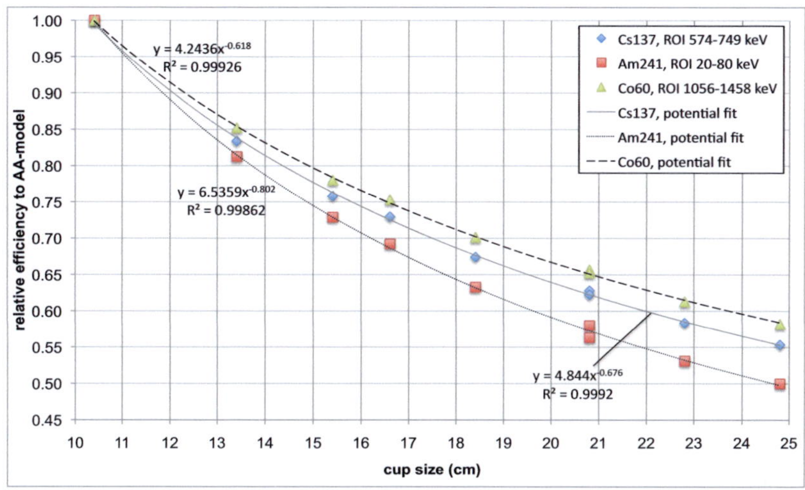

Figure 6.22: Detector counting efficiencies of the three radionuclides normalized relatively to the values of the AA-model plotted versus cup size (Hegenbart *et al.*, 2008).

Results and Discussion

For the low-energy range, glandularity seems to play a more important role, as can be seen at E40 for ^{241}Am (table 6.14). The difference of the E- and the E40-model is roughly 1.7% for ^{241}Am, while it is only 0.5% for ^{137}Cs and only 0.4% for ^{60}Co.

Higher glandularities will result in lower efficiencies. This is to be expected, as the mass and thus attenuation of the radiation increases. For smaller cup sizes than E the effect of glandularity is expected to be lower. This is due to the total breast mass being smaller and the fraction of glandular tissue (table 6.13) being closer to the median glandularity of 46.6% by Jamal *et al.* (2004) or of 40% by ICRP (2002). Therefore, deviations caused by variations in glandularity are negligible, because the woman's glandularity is usually unknown and difficult to determine.

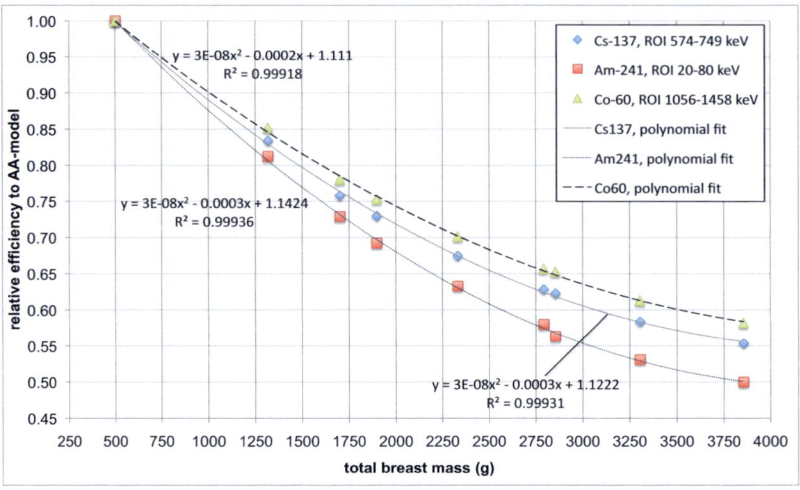

Figure 6.23: Detector counting efficiencies of the three radionuclides normalized relatively to the values of the AA-model plotted versus total breast mass (Hegenbart *et al.*, 2008).

Conclusion

In routine lung counting statistical counting errors are at least as high as 1.7%. The glandularity introduces a systematic error into activity measurement. However this error is negligibly small compared to other errors. For practical lung counting purposes, differences in breast glandularity can therefore be ignored.

6.3 Synopsis of Sensitivity Analyses and Measures

The research carried out in this chapter can be taken as example for real in-vivo measurement scenarios. The results for detector efficiency in individual cases may differ, but it is assumed that observed trends can be generalized and transferred to other cases. Table 6.15 summarizes the results of the research presented in this chapter and gives a rating for the sensitivity of each parameter.

Generally, the assumption was confirmed that low-energy photon emitters such as ^{241}Am are more sensitive for detector efficiency due to the higher attenuation coefficients in the low energy range. The only exception being medium-energy photons of ^{137}Cs, which resulted in higher sensitivity. This is due to the unfair comparison of the ROIs for the two nuclides, as the ROI for ^{241}Am included a wide scattering energy range.

Two organs have been examined: lung and liver. The obvious difference between the two are the density and the extent of asymmetry. The lung has a much lower density. When shrinking and organ displacements were considered, the lung showed a considerably higher sensitivity compared to the liver examinations due to the higher density of the fill material around the lungs. A lung has two lobes on each side of the body, while a liver is concentrated in the right body half. This is reflected more in absolute detector efficiencies than in relative values. No further differences were observed in the behavior of the two organs.

The studies showed that some parameters play a marginal role for detector efficiency for in-vivo scenarios. Inclination and rotations of the detectors have a low sensitivity. As well, segmentation errors or organ densities have a minor influence. While the examination table plays an important role when point-sources are simulated, this is not the case for volume sources when using voxel models.

Some parameters such as adipose/muscle tissue and voxel resolution of the model play a minor but not negligible role with regards to detector efficiency.

The position of the detector in relation to the source organ is a crucial factor. All parameters with changes in y-direction (anteroposterior direction) showed the highest sensitivity. Those were the detector- and organ displacements in anteroposterior direction, as well as the CWT and breast size.

With the results of the sensitivity analyses, measures have been developed and applied aiming to reduce systematic measurement errors of measurements. The next sections describe some of the successfully introduced measures.

Table 6.15: Summary of the sensitivity analyses expressed as simplified sensitivity ratings in three categories (low, medium, and high). If applicable, typical parameter ranges were assumed that occur in real measurements with real test persons. A low rating means that detector efficiency changes in the range of statistical counting uncertainties. A high rating means that the parameter is very sensitive and cannot be neglected. Medium ratings indicate a sensitivity between the two extremes.

Parameter description	Section	Parameter range	Sensitivity rating
single detector displacem.	6.1.1	± 1 cm in x-direction	high
		± 1 cm in y-direction	high
		± 1 cm in z-direction	medium
single detector rotation		$\pm 10°$	low
single detector inclination		$\pm 10°$	low
voxel-wise organ displacem.	6.1.2	± 1 voxel in x-direction	low
		± 1 voxel in y-direction	high
		± 1 voxel in z-direction	medium
chest wall thickness	6.1.4	± 10 mm	high
continuous organ displacem.	6.1.5	± 7 mm in x-direction	low
		± 7 mm in y-direction	high
		± 7 mm in z-direction	medium
lung shrinking	6.1.6	factor≈ 0.8	high
liver shrinking		factor≈ 0.8	medium
ratio adipose/muscle	6.2.1	0/100 to 87/13	medium
voxel size	6.2.2	2^3 to 6^3 mm^3	medium
examination table	6.2.3	level of detail	low
lung source distribution	6.2.4	volume / surface source	medium
anti-coincidence (^{241}Am)	6.2.5	on/off	low
lung segmentation errors	6.2.6	-25% of segm. volume	low
lung density	6.2.7	$\pm 10\%$	low
breast size	6.2.8	cup size AA to G	high
breast glandularity	6.2.9	7 to 40%	low

6.3.1   Positioning System

A result of the sensitivity analyses is that the relative position of the detectors to the subject has a considerable influence on detector efficiency. This quantity is hard to determine directly, but the position of the detector and the subject can be determined independently in an absolute coordinate system. A positioning system was designed and installed in the PBC chamber (see figure 6.24). In the following paragraphs, the installed system is described and how it is used for the determination of the phoswich detector position and the determination of the position of an object with landmarks.

All movable elements of the phoswich detector rack have been equipped with sensors from Heidenhain GmbH. Two Gage Chek computers from Metronics, Inc. display their measured data.

The sensor technology is based on an optical scanning of so called graduations, which are periodic structures created by photolithographic processes. A sensor head moves over a rail (length) or a disk (rotation) with graduations and sends electric pulses to the Gage Chek for each graduation. As the sensor head passes a reference mark, following graduations are incrementally counted. Passing of the reference marks sets the counts to zero.

Five rotary encoders[2] fitted with a disc of 3600 graduations have been installed to measure various angles on the phoswich detectors rack. The theoretical accuracy of these encoders is $0.1°$. The rotational parameters are displayed on the Gage Chek in the unit of degrees and arcminutes are the following:

W_{HA} is the angle of the main axis (German: *Winkel Hauptachse*).

W_{h1} is the angle of detector 1's horizontal axis (German: *Winkel der horizontalen Achse von Detektor 1*).

W_{h2} is the angle of detector 2's horizontal axis (German: *Winkel der horizontalen Achse von Detektor 2*).

W_{v1} is the angle of detector 1's vertical axis (German: *Winkel der vertikalen Achse von Detektor 1*).

---

[2]Type ERN1080

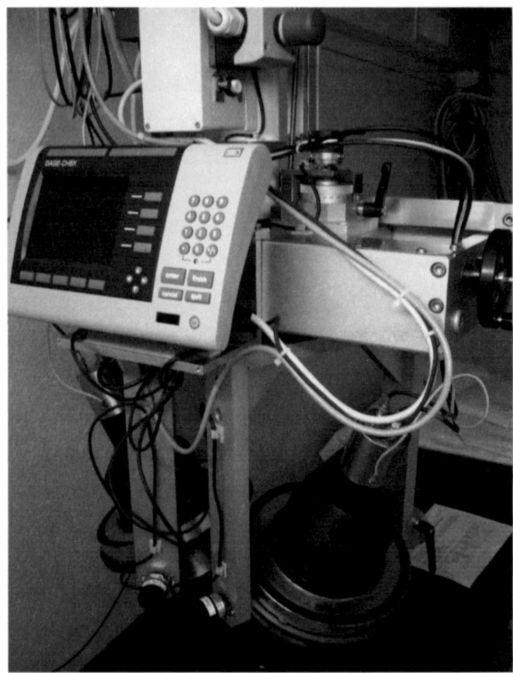

Figure 6.24: Gage Chek display mounted on the phoswich rack in the partial body counter chamber.

W_{v2} is the angle of detector 2's vertical axis (German: *Winkel vertikalen Achse von Detektor 2*).

A goniometer was used to determine an offset value in relation to the reference mark in order to calibrate each rotary sensors. The offset and the rotational sign of the value can be calculated by turning each axis to a known value using the goniometer. These offsets and rotational signs can be set on the Gage Chek's preferences.

Lengths are measured with different types of sensors, but all sensors are based on the principle mentioned above. Here, the sensor head travels over a rail with the graduations. The labeling of length parameters refers to a right-hand rule coordinate system introduced in the rectangular chamber. The origin of this system is on the floor in the north-west corner of the chamber. The x-direction is perpendicular from the west

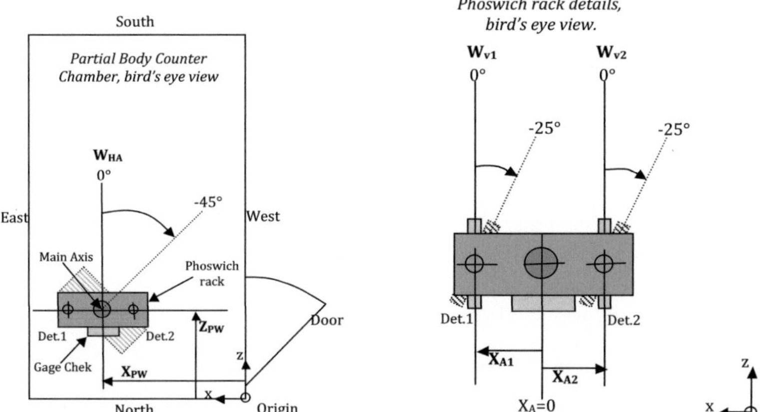

Figure 6.25: Left sketch: PBC chamber with the phoswich rack. This view clarifies the definition of the parameters W_{HA}, Z_{PW}, and X_{PW}. Right sketch: Details of the phoswich rack. This view clarifies the definition of the parameters W_{v1}, W_{v2}, X_{A1}, and X_{A2}.

wall to east wall. The y-direction is perpendicular from the floor to the ceiling, while the z-direction is perpendicular from the north wall to the south wall. The accuracy for all translational parameters is better than 1 mm. The Gage Cheks display the values in millimeters and tenths of a millimeter.

A sensor is used to measure the height Y_{PW} of the phoswich rack[3]. On the ceiling of the chamber another sensor measures the x-position X_{PW} of the main axis. The z-position Z_{PW} is measured analogous. An encapsulated graduation rail is equipped with two sensor heads measuring the distances X_{A1} and X_{A2} between the detector axes to the main axis. The label X was chosen because the vertical detector axes move in x-direction when the main axis is in $0°$-position.

Calibration was performed by determining the offsets for each length parameter. This was done by measuring the distance from the main axis to the corresponding point of interest, e.g. to the walls that represent a coordinates zero value, or to the two vertical detector axes.

---

[3] PW stands for phoswich rack

Figures 6.25 and 6.26 help to understand the definition and sign of rotation of all the ten parameters. Both Gage Chek computers can transmit the displayed values to the local network via a serial-interface-Ethernet-converter. With a virtual COM-port on any computer in the local network the values can be accessed online. Technical details, offset values, network preferences, routine procedures for calibration, and additional preferences are described in an internal laboratory report (Hegenbart, 2008).

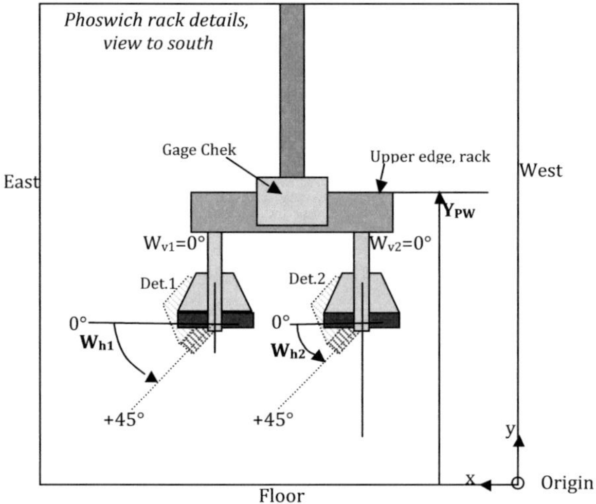

Figure 6.26: Side view of the phoswich rack that clarifies the definition of the parameters W_{h1}, W_{h2}, and Y_{PW}.

Finally the phoswich rack's main axis was equipped with a laser diode. Its laser beam is pointing vertically down to the floor and enables the user to locate of a point of interest. X_{PW} and Z_{PW} of the point can be read from the Gage Chek. To determine the height of the point of interest, a laser rangefinder is used. The rangefinder is attached to the phoswich rack. The height can be calculated by the displayed Y_{PW} minus the value read on the device. The manufacturer specifies an accuracy for the rangefinder of ± 1.5 mm.

Detector Positioning

The parameters measured by the positions system can be used for routine records and to set up Monte Carlo simulations of the very same measurement scenario. With the parameters defined in the previous paragraphs and some geometric constants, the detector position, i.e. both ($i = 1, 2$) detector axes' emission points p_i (the centers of the detector windows) and their vectors $\vec{v}_i$, can be determined unambiguously in the coordinate system of the room (see equations 6.6 and 6.7). Voxel2MCNP has an option to access the parameters from the Gage Chek via a local network and can position the virtual detectors in the program accordingly.

$$p_i = \begin{pmatrix} -a\sin(W_{hi})\cos(W_{vi}) + X_{PW} + X_{Ai}\cos(W_{HA}) \\ -a\cos(W_{hi}) + Y_{PW} - b \\ a\sin(W_{hi})\sin(W_{vi}) + Z_{PW} - X_{Ai}\sin(W_{HA}) \end{pmatrix} \tag{6.6}$$

$$\vec{v}_i = \begin{pmatrix} \sin(W_{hi})\sin(\frac{\pi}{2} + W_{HA} + W_{vi}) \\ \cos(W_{hi}) \\ \sin(W_{hi})\cos(\frac{\pi}{2} + W_{HA} + W_{vi}) \end{pmatrix} \tag{6.7}$$

with

$a = $ distance of horizontal detector axis to its center, and

$b = $ distance of horizontal axes to upper edge of phoswich rack.

A cylindrical detector model is coded in Voxel2MCNP. All detector information including geometric parameters, like radii and crystal thicknesses, are stored in the code. The software calculates p_i and $\vec{v}_i$ and generates the complete detector model in MCNP syntax. For cylindrical geometries the macrobody input card `rcc` is used to model the detector. The attributes of `rcc` are the components of p_i followed by the components of $\vec{v}_i$ in xyz-order, as well as the radius of the cylinders in centimeters (see also appendix C.3, page 213 for an example input file with `rcc` cards).

Object Positioning

In addition to the absolute position of the detectors in the chamber, the position of objects (e.g. subjects, phantoms, point sources, scattering items, ...) need to be known in order to set up a simulation. Landmarks need to be defined on the object. In the PBC chamber landmarks can be located with the laser beam and the rangefinder. To reproduce a landmark in the virtual world the determined coordinates can simply be entered in an MCNP input card. Alternatively the 3D-viewer of Voxel2MCNP (section 4.6.2) can be used. It displays a red line to picture the laser beam of the positioning system. Additionally a red disk can be moved up and down in y-direction along the beam to determine the height of a virtual point (figure 6.27) in relation to the coordinate origin in the PBC.

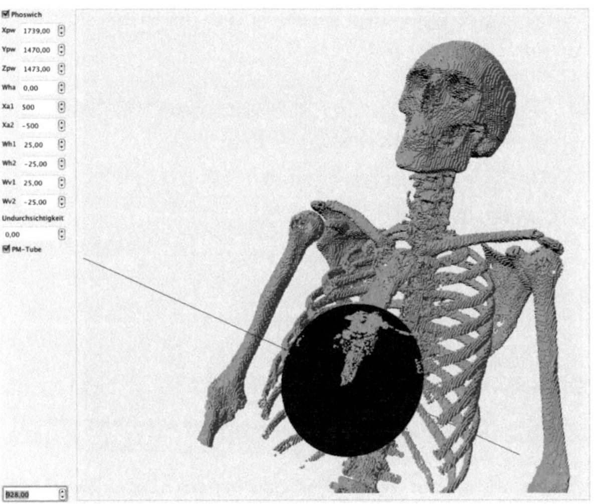

Figure 6.27: Voxel2MCNP's 3D-viewer enables the user to locate points of interest (e.g. the tip of the sternum) and displays the coordinates of that point.

To determine the position of radioactive point sources one landmark is enough, as they are small and point-like. This is under the assumption that their radiation field is isotropic. Generally a three-dimensional object's position can be determined distinctly by defining three landmarks. It is prudent to define more than three landmarks to compensate for errors of the positioning system and human error. For example, nineteen landmarks have been defined for the LLNL Torso Phantom (figure 4.20, section 4.7.8).

6.3.2 Hardware and Software

Monte Carlo simulations need computational power and if they include voxel models, large amounts of memory are required. The more complex the MCNPX input file, the higher the photon energy, and the larger the number of simulated particles, the longer the simulations take.

Section 4.7.7 described one way to reduce the number of voxels. In some cases however, highly resolved voxel models are necessary to generate accurate results, as concluded in section 6.2.2. Also as discussed in section 4.5.2, some radiation protection dose problems focus on parts of the human body that need a high resolution voxel model to be simulated accurately.

Computer Cluster

A cluster of computers was set up to facilitate the simulations. Simple office computer with two processor cores and 3 GB of RAM have been linked to a cluster in the network. The two cores of the office computers allowed simulations and normal office work at the same time without noticeable disturbance of the user. The precompiled MPI[4]-Windows-version of MCNPX 2.5.0 was installed on these computers with an MPICH.NT 1.2.4 client (Gropp *et al.*, 1996; Argonne National Laboratory, 2009b). This set up was used to allow parallel computing on several Windows computers at ISF. The cluster is extendable and new nodes can be added as required.

Using this set up, simulations can be executed in parallel to speed up the calculations by a factor almost equal the number of chosen processor cores. For example, a simulation of a lung counting scenario in the PBC with the MEET Man model with $(4 \text{ mm})^3$-voxels and with ^{241}Am as source was carried out with a single core on a single computer in 131 minutes. Using four additional cores of four additional computers, the same input file was completed in 28 minutes, thereby increasing the speed by a factor of 4.7.

The number of processors is limited. With the mentioned input file, the MPI-client reported a memory problem as soon as the number of

---

[4]Message Passing Interface

processors exceeded eight. It was observed, that the higher the number of voxels, the less processors could be used in a parallel calculation.

Recently, the ISF computer cluster was upgraded to the new version of MCNPX 2.6.0. It supports the more advanced standard MPI2 (Argonne National Laboratory, 2009a). The new version has not yet been tested to establish if the memory problem persists.

If parallel processing fails, the cluster still provides computing power to calculate a number of different simulations at the same time. This can then be controlled remotely from a single computer.

64-Bit Compilations

The office computers in the cluster at ISF are 32-bit systems. Own research showed that the 32-bit version of MCNPX can handle roughly up to 20×10^6 voxels. To address more than 4 GB of memory, a 64-bit computer system is necessary.

First tests with a self-compiled 64-bit version of MCNPX on an Apple MacPro computer with 64-bit Intel-Xeon processors and 16 GB of RAM showed that the number of voxels that can be handled by MCNPX can be increased dramatically. Practically the RAM of the computer sets physical limits. As soon as memory needs to be swapped to a hard drive, the code slows down and computing time increases beyond reasonable magnitudes.

The attempts at ISF on the MacPro computer showed, that the LLNL Torso Phantom in the fine resolution (about 90 million voxels) can be initialized with MCNPX 2.5.0. So far, an MPI version of the 64-bit compilation could not be built, so parallel calculation with other 64-bit systems is not possible, yet. Tests with the new MCNPX version 2.6.0 are ongoing. A new cluster based on 64-bit technology is scheduled.

6.3.3 Adapting the Model Geometry

Amongst the highest observed sensitivities are the position and size of the organs of interest. Ideally, the model used in the simulation should match the anatomy of the test person for the measurement. However, these details are hard to establish without the help of tomographic scans of the test person.

However, the outer shape of the test person is more simply to measure: anthropometric parameters like height, body weight, chest circumference, etc. are easy to determine. For instance, as described in section 6.2.8, it was possible to quantify the female breast size in bust and underbust girth to generate customized models with the BREP technology.

There are correlations of the inner anatomy and anthropometric shape parameters of humans. For example, Willmann *et al.* (2007) uses body weight and height to generate a population model including organ masses to assess the influence of individual variability on pharmacokinetics of drugs.

The following chapter 7 discusses the difficulties in creating customized models and describes a new method to generate adapted models by image registration algorithms.

Chapter 7

Customized Models

Every person is an individual. Anthropometric values like body weight, height, lung volume, breast size, chest wall thickness, or chest circumference differ from person to person. The distribution of tissues like bones, adipose, muscles, or soft tissue in the body is also specific to each individual. The individual also changes over time. Long term changes are caused by age. The human body changes from childhood, adolescence and adulthood to seniority. Mid term changes are caused for instance by nutrition or plastic surgery. Short term changes are caused by for example breathing, digestion, or posture. All these parameters have an influence on radiation attenuation. This chapter discusses the possibilities for creating customized models and presents a successful approach to transform models to match an individual better than a standard calibration phantom.

7.1   Segmentation

Segmentation of an individual's tomographic data can be used to create a customized model. Photographs of body slices, as those taken for the Visible Human Project (Ackerman, 1991), can only be taken postmortem. Therefore they are not feasible for living subjects. The advantage of tomographic data is that the segmented models represent the individual one-to-one, as tomographic data represents the posture of the subject when the data was taken. Short term changes are not taken into consideration. Tomographic data of an individual can be obtained using CT and MRT technology. The feasibility of both technologies for creating customized models are briefly discussed in the following subsections.

Computed Tomography

CT is an excellent method to locate bones, lungs tissue, soft tissue, and air inside the human body. The mentioned organs are relatively easy to segment. It is more difficult to distinguish between different soft tissues.

Segmentation of relevant organs for in-vivo diagnostics can be performed with modern semi-automatic segmentation algorithms[1], good anatomical knowledge, and extensive experience. An example for a segmentation of a CT-scan (of the simple anatomy of the LLNL Torso Phantom) is described in details in section 4.7.1. The purpose of in-vivo bioassay is radiation protection. Generally, it is ethically, and legally not justifiable to take CT scans of persons without medical indication. Therefore segmentation of CT scans is not a feasible method.

Magnetic Resonance Tomography

The advantage of MRT is that it operates with non-ionizing radiation, hence there is no pathogenic radiation dose. Soft tissue can be better differentiated by MRT than with CT. MRT is more expensive and takes more time compared to CT.

A high hydrogen content of tissue causes an intense signal (e.g. fatty tissues or water depositions). Calciferous hard bones give a weak signal due to small amounts of hydrogen. The contrast of tissues, e.g. arteries and veins, can be improved by applying tracers.

Patients with pacemakers, vascular clips, or protheses might have difficulties getting MRT scans. This is due to dislocation of metal objects by magnetic forces or induction of electric currents that could harm the patient or destroy the item.

---

[1]provided by tools such as OsiriX (The OsiriX Foundation, 2009) or the Insight Toolkit (Kitware, Inc., 2009c)

The IVM would need periodic support of facilities that provide tomographic scanners whenever a scan is needed to create an individual model. Still, segmentation of tomographic scans is a tedious, time-consuming procedure. Therefore segmentation of tomographic scans is not feasible for the day to day work at IVM. It was decided not use segmentation techniques to create customized models of subjects.

7.2 Image Registration

7.2.1 Basic Principles

Image registration is the process of transforming at least two sets of data into one coordinate system. Classically, image registration is necessary in order to be able to compare or integrate the data obtained from different measurements of the same subject. For example, medical image data of the same patient is taken by different tomographic modalities. Then, image registration is used to merge all data into one visualization. With this technique, a CT and PET[2] images are superimposed, so that areas of abnormality on the PET image can be correlated to anatomical details on the CT image (Yoo, 2004).

An image data set that needs to be transformed is called a *moving image*. Another image data set acts as the template for the transformation, which is fixed. Therefore, this is called the *fixed image*. Registration is an optimization problem. The aim is to find the optimal transformation that delivers the best match of the moving with the fixed image.

7.2.2 Concept for Customized Models

Image registration techniques can be used to create customized models from already segmented voxel models, i.e. using segmented voxel models as moving images. The fixed images originate from individuals. If already available voxel models can be transformed to match the individual, timely segmentation can be avoided. In section 7.1 it was concluded that segmentation of tomographically acquired data is not feasible for the

---

[2]Positron Emission Tomography

routine work at IVM. Therefore, a practical approach to create individual models based on registration techniques was favored.

The simplest way of transforming an already segmented voxel model is to stretch it linearly in all three dimensions to match the height, width, and depth of an individual. Often the body proportions of the individual and the available voxel model are not similar. Therefore, more advanced transformation techniques are necessary. The approach discussed in this section, allows gathering of more anthropometric data than just body weight and height.

As discussed, the fixed image must be generated by other means than tomographic scans. A whole body scanner, which can be based on laser range-finding or light-section methods, can generate a three-dimensional point cloud of a test person's body surface. The test person should be dressed only in underwear in order to exhibit the body proportions. Ideally, the test person should be scanned in a similar posture as the voxel model that is planned to be used as a moving image. Since the in-vivo measurements are taken in supine position and due to gravity effects on the anatomy, it would be best to scan the test person's body surface in the same position, right before the measurement on the examination table. Likewise, the moving image voxel model should represent a person in supine position, too.

In this way, the point cloud of the test person can then be used as fixed image data. The segmented voxel model's body surface can be used as moving image data. Then the transformation resulting from the registration procedure is applied to the complete voxel model to create a customized voxel model.

This however, does not address the problem that a whole body scanner cannot scan the inside of the body. Therefore, the position of organs and the thickness of adipose layers as well as chest wall cannot be determined. Nevertheless, the introduced concept provides options to improve the generation of customized models rather than linearly scale the voxel models. This approach will be further discussed in the following subsections.

7.2.3 Methodology

In the following paragraphs, the basic methods of image registration used in this research are introduced briefly.

Iterative Closest Point Algorithm

The Iterative Closest Point (ICP) algorithm (Besl and McKay, 1992; Ibanez *et al.*, 2005) is employed to match two point clouds. The ICP algorithm is often used to reconstruct 3D surfaces from different objects. The algorithm is simple and iteratively estimates the transformation between two point clouds. Essentially the algorithm's iteration loop consists of four steps. Firstly, the points are associated by the nearest neighbor criteria. Secondly, the transformation parameters are estimated by using a cost function. Thirdly, the points are transformed according the estimated parameters. Finally, the points are re-associated and a new iteration is started. The advantage of ICP algorithm is that it is applicable without prior knowledge of the corresponding points of the two point sets. Therefore the number of points of the two point clouds may vary. The Insight Toolkit 3.4.0 (ITK[3]) provides a software implementation of the ICP algorithm as an point-set-to-point-set registration method.

Rigid Transformations

There are two types of registration, rigid and non-rigid (elastic) registration. Usually registration is performed in different consecutive levels. The first step is a coarse registration that roughly aligns the moving image to the fixed image. This is mostly a rigid registration. Once the images are roughly aligned, further, finer registrations can be performed. The simplest rigid transformations for registration are translation and rotation. Both are merged together in one ITK method called `itkEuler3DTransform`.

Another popular and more advanced transformation is affine transformation. Geometrically, affine transformation transforms a cube into a hexahedron. It preserves the collinearity relation between points. For

---

[3]The Insight Toolkit (Kitware, Inc., 2009c)

example, three points on a line are also on a line after the transformation. Also the ratios of the distances of the points to each other are preserved. An affine transform combines a number of single linear transformations (rotation, scaling or shear). In addition a translation $\vec{b}$ accompanies the affine transformation as defined in equation 7.1.

$$\vec{x'} = A\vec{x} + \vec{b}$$

$$\begin{pmatrix} x'_1 \\ x'_2 \\ x'_3 \end{pmatrix} = \begin{pmatrix} a_{1,1} & a_{1,2} & a_{1,3} \\ a_{2,1} & a_{2,2} & a_{2,3} \\ a_{3,1} & a_{3,2} & a_{3,3} \end{pmatrix} \cdot \begin{pmatrix} x_1 \\ x_2 \\ x_3 \end{pmatrix} + \begin{pmatrix} b_1 \\ b_2 \\ b_3 \end{pmatrix} \qquad (7.1)$$

$$A' = \begin{pmatrix} a_{1,1} & a_{1,2} & a_{1,3} & b_1 \\ a_{2,1} & a_{2,2} & a_{2,3} & b_2 \\ a_{3,1} & a_{3,2} & a_{3,3} & b_3 \\ 0 & 0 & 0 & 1 \end{pmatrix} \qquad (7.2)$$

ITK's implementation of the affine transformation is called `itk-Affine-Transform`. The transformation is represented by the 4×4 matrix A' (equation 7.2).

Deformable Transformations

Non-rigid registration involves a deformation of the image data. This might be necessary due to different breathing phases, different postures, or other anatomical changes in the image data sets available. A paper by Rueckert *et al.* (1999) describes a method of a free-form transformation based on B-splines used for non-rigid registration. Control points are defined on a superimposed grid on the image. Deformation coefficients assigned to the control points have a local deformation effect.

ITK provides a B-spline transformation that can be combined with the ICP registration algorithm, but there are a couple of unresolved issues. As proposed by Rueckert *et al.* (1999), the ITK implementation also uses a B-spline coefficient grid for the free-form deformation of an image or point set.

There are drawbacks to the ICP implementation with B-spline transformation. The number of coefficients is high. It is depending on the order of B-spline type (usually 3rd-order) used, the dimension (three for Euclidean space), and the resolution of the deformation grid. As a minimum, the grid therefore needs two control points and three (in case of 3rd-order B-spline) support points in each dimension. Hence the most basic setup for the B-spline transformation has $3 \times 5^3 = 375$ coefficients. The control point grid resolution needs to be in the magnitude of the deformations to achieve an accurate result. The number of control points in the image would just be $2^3 = 8$ in this example. This is not sufficient to accurately transform a voxel model. A large number of coefficients is needed which generates high demands on computing power.

The number of points in the point clouds needs to be higher than the number of coefficients to make a calculation possible, otherwise the problem would be under-determined. It is possible to reduce the number of points of the point cloud close to the number of coefficients to reduce computer time. This was tested, but the tests have been cancelled inconclusively after days of computing time. Simplified tests with small numbers of control points did not converge and coefficient values could not be constrained. Due to the described difficulties and time constraints of the project, the approach of non-rigid deformation was abandoned.

7.2.4   Implementation

The implementation of the proposed method is part of the Voxel2MCNP code (section 4.6) to create customized models with an ICP algorithm with the classes available from ITK. The procedures described here can be can be intialized from the GUI[4]. A flowchart (figure 7.1) outlines the different steps.

---

[4]Graphical User Interface

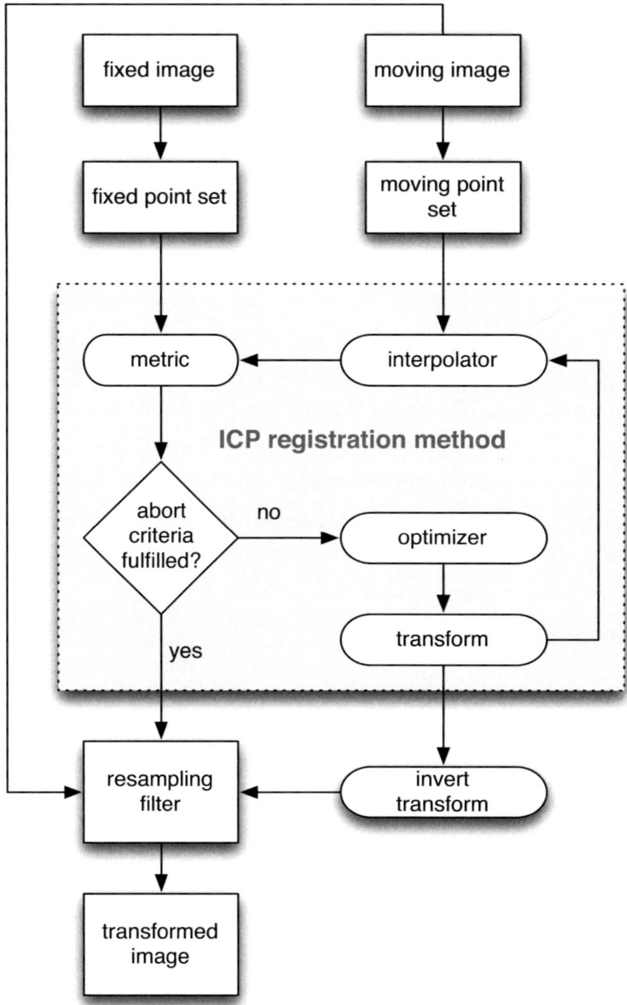

Figure 7.1: The flowchart illustrates, how the proposed registration method is applied to transform a voxel model (moving image) to match a fixed image.

Creation of a Point Cloud

The first step is to create point clouds (i.e. point sets) of the fixed and the moving image. The subject's surface point cloud can be taken from a 3D whole body scanner. To create the point cloud from the loaded voxel model some basic steps have to be performed first.

To extract the voxel model's surface the outside and inside of the model's volume must be clearly separated. Usually air voxels define the outside volume. If model body's inside contains air, e.g., in the trachea or intestine, this air needs to be redefined as a new voxel type, for example *air inside the phantom*. This enables Voxel2MCNP's organ isolation feature to define two voxel types: voxels in- and outside the model body. Voxel2MCNP's region growing feature can be used to fill the air inside the model. It may become necessary to manually draw a border between the inside and the outside air, to prevent flooding of the outside air segment. Once internal and external volumes are clearly defined, all organs can be merged into one voxel type. Boundary voxels can now be determined and their centers can be exported as a point cloud of the body's surface and saved as an ASCII file. Voxel2MCNP saves the point's coordinates temporary and for future use. The voxel model can be coarsened to reduce the number of points in the cloud and to speed up the registration.

Execution of the Registration

From the GUI the Euler-3D- or Affine transformation can be selected for registration. The user chooses the point cloud files for the fixed and moving image. The point clouds are stored memory in ITK's point set containers. ITK's point-set-to-point-set registration method is initialized and accompanied by an Euclidean distance metric and by a Levenberg-Marquardt optimizer. Depending on the number of points in the point sets, the chosen transformation and the quality of the initial alignment, the algorithm will take between a couple of minutes and a few hours to complete. Finally, the registration outputs the parameters of the chosen transformation.

Afterwards the Euler-3D- or the Affine transformation is inverted with ITK's `GetInverse` function. This is necessary, because of two methodologies used in ITK: forward and backward warping[5]. In this implementation both warping principles are used. The ICP point set registration method is based on forward warping while the image registration methods, which are used to resample the image, are based on backward warping.

The inverted transformation is then applied to the loaded voxel model. In order to do this, the loaded voxel model is first defined as the moving image in ITK's image class. When a transformation is applied to a voxel image, it is most likely that the voxels are moved to non-grid spaces. Therefore, remapping of the transformed model becomes necessary. This is done utilizing a resampling image filter in combination with a nearest neighbor interpolator. Other interpolators cannot be used for resampling segmented voxel models, because applying other interpolators could create undefined new organs. The organ identification values are discrete values, which cannot be interpolated with continuous *average* values.

An example for the registration is illustrated in figures 7.2 and 7.3.

---

[5]For details see section 8.3.1 *Direction of the Transform Mapping* of the ITK Software Guide (Ibanez *et al.*, 2005)

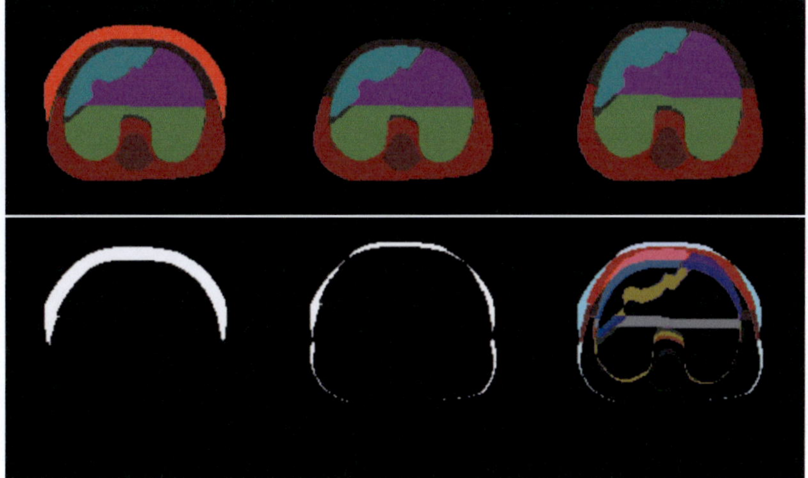

Figure 7.2: The upper three images show sections of the model of the LLNL Torso Phantom with overlay 4 (upper left image) which was used as the fixed image, the LLNL Torso Phantom without overlay which was used as the moving image (upper middle image) and the affine registered (transformed) model (upper right image). The lower three images are differential images generated from the upper images with the GIMP. The lower left image shows the difference of non-air voxels between moving and fixed image, resulting in the overlay 4. The lower middle image shows the difference of non-air voxels of the fixed image and the transformed image. The lower right image is a colored version of the differential image of fixed and transformed image. It illustrates considerable shifts of the organs in the registered model.

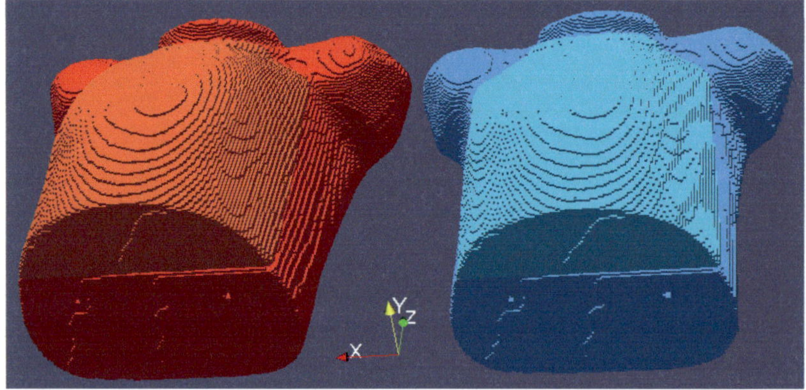

Figure 7.3: The red model on the left hand side shows the model transformed after an affine registration of the model without overlay (right turquoise model) into the model with overlay 4. The transformation stretched the phantom in mainly in y-direction. The matrix of the transformation for this example according to equation 7.2 was:

$$A' = \begin{pmatrix} 1.045 & -0.002 & 0.001 & -11.091 \\ -0.002 & 1.107 & -0.015 & -11.525 \\ 0.001 & 0.002 & 0.995 & 0.330 \\ 0 & 0 & 0 & 1 \end{pmatrix}.$$

7.2.5 Creation of Customized Model

Monte Carlo investigations have been carried out to determine if the proposed approach of generating customized models with rigid registration is useful to improve detector efficiency calibration. Voxel models have been registered with the discussed methodologies and were used to determine the detector efficiency for lung and liver counting with MCNPX. ^{241}Am has been used as a representative for low-energy photon emitters. Then, efficiencies have been compared with those determined by the current method at IVM (section 3.3.2).

Three model pairs have been chosen according to their chest wall thicknesses (CWT). The CWTs of the LLNL Torso Phantom with and without overlays are given by the manufacturer. These represent the moving images. A voxelized representation of the RPI Male Adult, the Zubal torso voxel model and the MEET Man $(3\ mm)^3$ voxel model (all introduced in section 4.5) were used as fixed images, hence these are the virtual test persons for this experiment. Table 7.1 shows the CWTs of various models calculated using equation 3.4. This equation is used for all test persons for lung or liver measurement at IVM. CWT of the MEET Man and the Zubal torso model was calculated using height and body mass. The RPI Male Adult (Xu *et al.*, 2008) resembles the ICRP Reference Man, hence the CWT was calculated using the data from ICRP (ICRP, 2002). The models of the LLNL Torso Phantom with the closest CWT are the moving images. As a result the model of the LLNL Torso Phantom with overlay 3 was chosen as a moving image for the MEET Man and the model with overlay 2 was selected for the Zubal torso model and the RPI Male Adult.

Point Clouds and Registration

As described in previous paragraphs, five models were utilized in the registration experiments. The model of the LLNL Torso Phantom with overlay 2 and 3, the MEET Man, the RPI Male Adult and the Zubal Torso Phantom. The last three models had to be cut in order to match the anatomy of the LLNL Torso Phantom. This was done with Voxel2MCNP's cutting feature. The transversal plane of the cut was determined by orientation on the lumbar and cervical vertebrae. The MEET Man's and

Table 7.1: The CWT of different models sorted by CWT.

Model	CWT (mm)	Mass (kg)	Height (cm)
LLNL Torso Phantom	12.8	n/a	n/a
LLNL Torso Phantom w/ overlay 1	18.4	n/a	n/a
Zubal Torso Phantom	22.5	70.3	177.8
ICRP Reference Man	23.4	73.0	176.0
RPI Male Adult	23.4	73.0	176.0
LLNL Torso Phantom w/ overlay 2	24.5	n/a	n/a
MEET Man	27.8	92.0	180.0
LLNL Torso Phantom w/ overlay 3	28.8	n/a	n/a
LLNL Torso Phantom w/ overlay 4	35.8	n/a	n/a

RPI Male Adult's arms were cut just below the shoulders (figure 7.4). A space of approximately 10 cm was added around each model to give the model enough space for transformations.

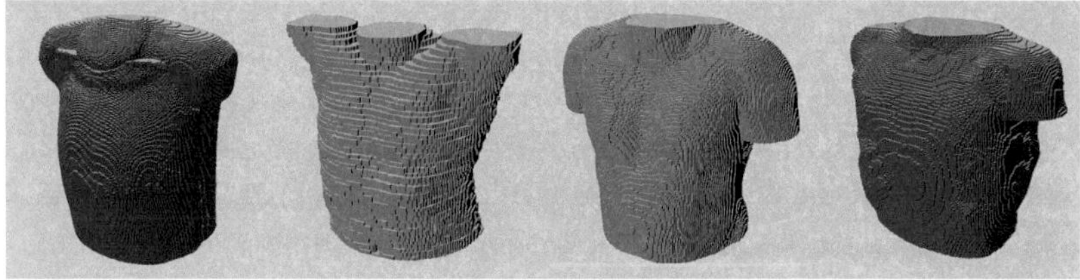

Figure 7.4: Torso models of (from left to right) the LLNL Torso Phantom with overlay 2, the Zubal torso phantom (head and lower abdominal part cut), the MEET Man, and RPI Male Adult (lower abdominal part including legs, arms, and head cut).

Air inside the LLNL Torso Phantom and the MEET Man was redefined to distinguish between the model body's inside and outside. The points clouds had been extracted from the five models and saved in an ASCII file.

The registration of the point sets was performed in two steps. First a 3D-Euler registration, then an affine registration was carried out. The transformation had been applied to the LLNL Torso Phantom with overlay 2 and overlay 3, respectively.

Monte Carlo Simulations

Monte Carlo simulations with MCNPX had been performed. The setup was according to the standard lung or liver measurement used at the IVM with the phoswich system. This means, detectors are inclined and rotated by $25°$ and positioned directly over the skin of the model above the organ of interest. First the LLNL Torso Phantoms without and with all four overlays have been simulated to generate data for a calibration curve (detector efficiency over CWT) for lung and liver measurement. Microsoft's Excel was used to plot the data and calculate a fit curve by linear regression.

Afterwards the MEET Man torso, the RPI Adult Male torso, the Zubal torso, and the transformed LLNL Torso Phantoms have been simulated with the same setup. The detectors' position was kept constant for all measurements, except the height that was adjusted to be about 1 mm over the model's surface. The ROI for ^{241}Am was chosen from 20 to 80 keV. Detector efficiencies of both phoswich detectors were summed up. The number of simulated particle histories was 10^7 for each simulation.

Results

Both, lung and liver measurements show relatively similar results, there-fore, they are not discussed separately. The calibration curves generated from five data points had very good R^2-values of greater than 0.99. The exact values are displayed in figure 7.5. The liver and lung detector efficiency calculated for the MEET Man and the Zubal torso model are considerable lower than the corresponding values on the calibration curve. This illustrates the large systematic errors that can occur with the current calibration methods used at IVM.

For the lung (liver) simulations, the calibration curve gives a 25% (75%) higher detector efficiency than the simulated efficiency for the MEET Man. The registered model gives an overestimation of just 14% (62%). The results for the Zubal torso model are similar. The calibration curve gives a 35% (82%) higher detector efficiency than the simulated detector efficiency. The registered model for the Zubal model gives an overestimation of just 11% (64%). The results of the RPI Adult

Male are closer to the calibration curve. The calibration curve gives am overestimation of 14% (4%). The corresponding registered model gives an overestimation of just 5% (2%). The detector efficiencies for RPI Adult Male are already close to the chosen moving image model with overlay 2, hence the smallest absolute improvements are observed here.

Conclusion

Three models, the MEET Man, Zubal, and RPI Adult Male were used as fixed images with a novel registration technique for voxel models. For all presented cases, considerable improvements could be achieved. The calculated values for the registered models are closer to the correct values of the original models than the values of the calibration curve.

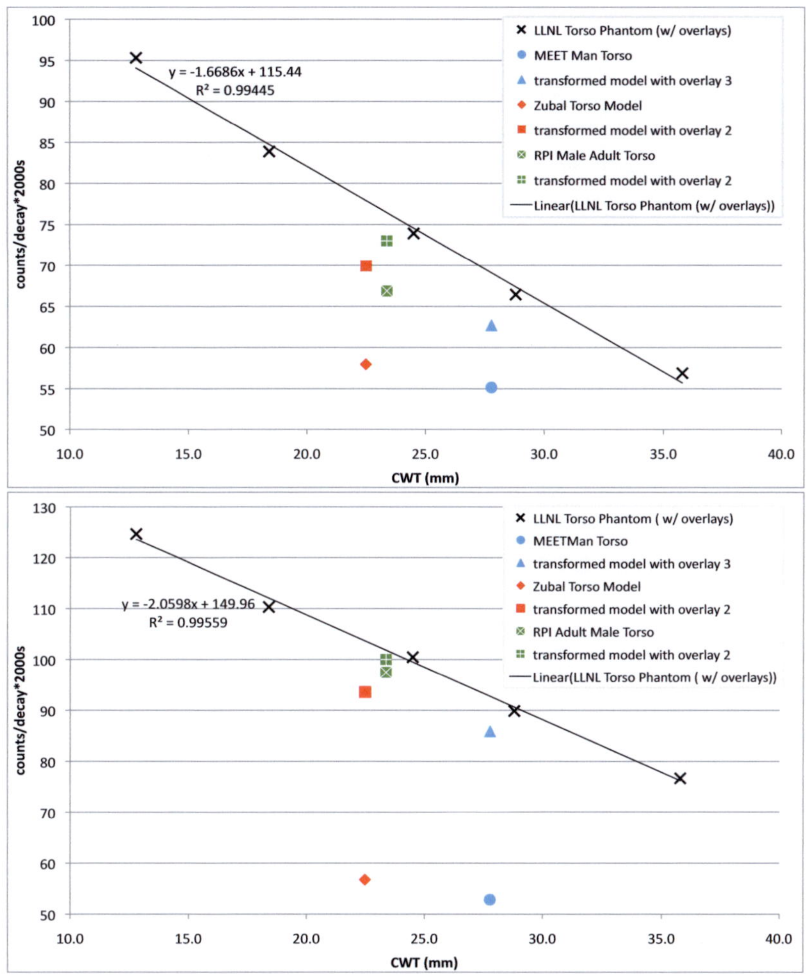

Figure 7.5: Lung (upper images) and liver (lower image) counting detector efficiencies plotted over CWT. The calibration curve was created by simulation of different overlays of the LLNL Torso Phantom (black crosses). The MEET Man Torso, the Zubal Torso Phantom, the RPI Adult Male Torso and the transformed models are colored.

7.3 Summary and Outlook

Finding an appropriate method to generate customized human models for day to day use at IVM presents certain challenges. Customized modeling derived by segmenting tomographic data is possible, as demonstrated with a simple phantom (section 4.7.1). In addition to previously discussed issues, segmentation is very time consuming. Segmentation is therefore not suitable for routine applications, as results for a suspected incorporation of radionuclides are expected promptly, allowing physicians to quickly take action and treat the test person appropriately.

The approach presented in this thesis is based on image registration and shows promising results, although only three voxel models have been examined. Utilizing deformable registration techniques could improve the shape of the transformed moving image and therefore presents opportunities for further development. Other possible areas of future research could focus on the internal anatomy of the subject. A mass-spring-damper model, an approach presented by Jarrousse *et al.* (2008), for instance, can be used to transform the inside of the moving image in a controlled way to generate a desired CWT.

In order to do this, additional data must be gathered. Body height and weight would not be sufficient. Research into the use of an ultrasound device to determine CWT could be undertaken. Spirometry could be used to gather volume data of the lung tissue. Measuring adiposity could also improve the achieved results. Adiposity of subjects can be estimated in a number of different ways. One method is based on the determination of ^{40}K in the body that can be measured in the WBC at IVM.

Chapter 8

Conclusion and Perspective

8.1 Problems Recapitulated

In vivo measurements are performed to measure incorporated radionuclides within the human body. To determine the activity in the test person and to estimate the absorbed dose, the knowledge of efficiency of the detectors is of utmost importance. While statistical uncertainties can be reduced by increasing measurement time, using more detectors or increasing their size, there are also systematic errors that are difficult to control. Detector efficiency calibration depends on a number of different parameters. The research on the importance of these parameters in the laboratory environment is challenging, because the possibilities of measurements are limited. Monte Carlo methods therefore became the state of the art for complex radiation transport problems. Computational power of today's computers allows the comprehensive examination of various parameters. Even unrealistic scenarios can be analyzed.

To do this, virtual models of detectors and equipment have to be created and validated. The purpose of this work was to design a valid simulation environment for in vivo measurements. Appropriate human models are needed for the improvement of existing efficiency calibration methods. Ideally, there is an individual model for every subject.

Anthropomorphic phantoms are used for classical efficiency calibration. Such phantoms are not very flexible. Individualization is limited in terms of available radionuclides, postures, breast sizes, or organ variations. Numerical efficiency calibration based on Monte Carlo methods use virtual human models. These are stylized, voxel, or mesh-based models, which can be modified more or less easily, depending on the type. Voxel models provide the most realistic representation of the human body, since they are mostly based on tomographic scans.

8.2   Presented Solutions

Managing voxel models is a task that exceeds human abilities due to the large number single voxels, hence the application of information technology is inevitable. A software called Voxel2MCNP was developed to handle voxel models. This tool enables the user to generate entire virtual in vivo measurement scenarios at IVM, including detectors, equipment, chamber, and of course voxel models. The scenario can be visualized in 3D. The tool generates an input file for a wide-spread Monte Carlo particle transport code called MCNPX. The results of MCNPX simulations can be exported in form of tables and evaluated within spreadsheets. Detector efficiencies can be calculated this way. Voxel2MCNP became established at the ISF and is a valuable tool that many scientists and students regularly use for studies with voxel models.

A new voxel model from the LLNL Torso Phantom, which is a common anthropomorphic calibration phantom, was generated based on CT scans. The model is so far the most detailed voxel model of this phantom available. All physical separable parts are segmented and can be defined as a radioactive source or can be modified for specific examinations.

There was a need to validate the simulation environment, including models and generated scenarios to ensure that simulations would reflect real measurements. The new voxel model of the torso phantom enabled the comparison of measurements and simulations. The models of the WBC and the NaI(Tl)-detectors, the PBC including the phoswich pair and one HPGe detector have been validated with various radioactive point sources or the voxel model of the LLNL Torso Phantom.

Various parameters have been analyzed for their sensitivity on detector efficiency. The most sensitive parameters have been identified and quantified for a specific scenario. The results show a high sensitivity for the distance between detector and the subject's source organ. Generally, the detector-subject position plays an important role, leading to the decision to design and build a positioning system in the PBC chamber. The system displays the position of the phoswich pair. It is also able to determine the position of objects like point sources or phantoms in the chamber.

Some of the effects of source distribution, scattering, and photon energy have been examined. Valuable conclusions could be drawn from Monte Carlo simulation results. This provided a better understanding of the effects that could not be discussed utilizing analytical mathematics.

Another outcome of the research was that using a highly resolved voxel model provides more accurate results. Simulations with high-resolution models cause a high demand on computational power. A computer cluster was set up to enable his process. It is composed of regular office computers and is used to speed up calculations and to facilitate research. Students and scientists at the ISF make substantial use of the cluster, which helps to perform Monte Carlo simulations quicker.

Other parameters like CWT, breast size, and organ position also showed high sensitivity on detector efficiency. With deformable mesh-based models, it was possible to generate a series of individual female voxel models with different breast sizes and different glandularity of the breast tissue.

In addition to the approach with mesh-based models, a new method based on image registration was introduced to design customized models. The method enables to modify an already segmented model to match the characteristics of an individual. It is controlled by the outer contours of the test person that can be recorded by a whole body scanner. Virtual experiments, supported by new valid simulation environment, showed improvements of the classic efficiency calibration method. Using transformed models based on the image registration technology in Monte Carlo simulations for the determination of the detector efficiency resulted in smaller systematic errors.

8.3 Perspectives

The results and solutions presented in this work are promising and give reason for continued research and development in this field. The following paragraphs briefly outline areas of interest for further study.

The **segmentation** of the LLNL Torso Phantom, lead to a better understanding of this area. As a result of this work the ISF has now the facilities to segment other (calibration) phantoms, which creates opportunities for further studies.

The new **positioning system** in the PBC is connected to the network. The parameters send by the Gage Chek could be used for quality assurance purposes that can be noted in the subject's records.

It is already planned to expand the **computer cluster** with new hardware. The software for the cluster could be extended by sophisticated load balancing systems, user- and priority management systems. This would provide a continuous, fluent workflow of Monte Carlo simulations at the institute. Scientists and students at ISF and therefore concentrate on research rather and have peace of mind that any administrative tasks linked to simulations are dealt with by the cluster software.

The newly developed software **Voxel2MCNP** can be extended easily, since it is coded in a object-oriented manner. Manifest improvements would be the implementation of more detector models and equipment objects for a greater variety. A coordinate display for the mouse pointer in the 2D- or 3D-viewers would aid the navigation through the voxel model and the entire process. To choose and add objects, drag and drop or interactively manipulate objects in the viewer, would allow to quickly set up complex scenarios.

An further important improvement would be to generate *input files* or macro codes for **other Monte Carlo codes** like Geant4 or EGSnrc, which offer features that are not available in MCNPX. Furthermore, it is wise to confirm unexpected results with a second, independent code.

For gamma-spectrometry purposes an automatic **spectrum analysis** feature would be useful to speed up data evaluation of the Monte Carlo simulation results. A retroactive Gaussian broadening or a spectrum summing function for radionuclide mixtures are just two examples.

It is the intention to write a detailed **developer's guide** for Voxel2-

MCNP to encourage future users to further develop Voxel2MCNP at ISF. Another option would be to provide Voxel2MCNP to the open-source software community to involve an international community.

Voxel2MCNP should not be restricted to voxel models. The support of **mesh-based** or stylized model formats would broaden the possibilities of application.

It may be possible to link the software to **external nuclide libraries** available on the internet to *download* desired spectra for photons or electrons.

Improvements in **customizing models** have already briefly been mentioned in section 7.3. Problems experienced during the development of the application of deformable registration techniques still need to be resolved. This may be achieved by coding a new constrainable, deformable registration algorithm, in order to overcome the deficiencies of the ITK implementation. This would improve the shape of the transformed moving image.

The requirements for the acquisition of a reasonably priced **whole body scanner** for test persons should be evaluated. Technical and practical constraints, such as the space requirement in the measurement chamber, must be considered.

Future research will doubtlessly be focussed on the internal anatomy of the subject. For this, more anthropometric information is needed than just body height and weight. The possibilities of measuring **CWT** with ultrasound could be tested using modern ultrasound techniques. Another area of interest is the methodology used to determine CWT on voxel models. Clearly defined and reproducible methods must be developed for this.

Spirometry could be used to obtain **lung volume** data. A method of measuring **adiposity** based on the determination of ^{40}K in the body could be developed. The WBC at IVM is able to measure the ^{40}K-activity per kilogram, which gives indications on the adiposity of the test person.

In order to built a customized model based on this new data, a **mass-spring-damper** model could be used to transform the inside of an existing voxel model in a controlled way. Alternatively, **deformable mesh-based models** could be used, provided plausibility test are carried out on their anatomy.

Appendix A

Physical Fundamentals

A.1 Decay and Activity

Radioactive sources used in experiments throughout this work are subject to decay. The activity $A(\tau)$ at the time τ of a pure radioactive source is halved after its half-life $T_{1/2}$ and so is the number of atoms $N(\tau)$. The sources are usually shipped with certificates declaring the initial activity A_0 (of initial N_0 atoms) and its error, i.e. usually a single standard deviation σ. The activity during an experiment was calculated with equation A.1. The sources used have long half-lifes, therefore using a designated accuracy of τ in days is sufficient.

$$A(\tau) = A_0 \cdot e^{-\frac{\ln 2 \cdot \tau}{T_{1/2}}} \tag{A.1}$$

$$N(\tau) = N_0 \cdot e^{-\frac{\ln 2 \cdot \tau}{T_{1/2}}} \tag{A.2}$$

$$A(\tau) = N(\tau) \cdot \frac{\ln 2}{T_{1/2}} \tag{A.3}$$

A.2   Photon Attenuation

Also the attenuation is an exponential process and it is described by equation A.4. The intensity I of an X-ray or gamma ray at the distance x from the source, where the intensity amounts I_0. The attenuation coefficient μ is the sum of attenuation coefficients for each interaction of the photon with matter, e.g. photo-effect, compton-scattering, or pair-production. The ratio $I(x)/I_0$ is called transmission.

$$I(x) = I_0 \cdot e^{-\mu \cdot x} \tag{A.4}$$

with

$$\mu = \mu_{photo} + \mu_{compton} + \mu_{pair} + \cdots \tag{A.5}$$

Appendix B

Used Software Tools

B.1 Applications

For this thesis a comprehensive use of various software was necessary. This small section describes shortly the five most important software tools and their tasks, for which they have been used in this work:

Paraview in version 3.4.0 (Kitware, Inc., 2009b) to view three-dimensional images of various formats, for example VTK[1], VRML[2].

OsiriX in versions from 2.6 to 3.2.1 (The OsiriX Foundation, 2009) to view DICOM files and segmentation of CT scans.

Microsoft's Excel in version 12.1.5 (Microsoft Corporation, 2009a) for spread sheet calculations and to plot graphs and fit curves.

Gnuplot in version 4.2 (Williams *et al.*, 2009) to fit and plot curves.

GIMP (GNU[3] Image Manipulation Program) in version 2.4.7 (The GIMP Team, 2009) to manually manipulate voxel model slices and to convert file formats from TIF to BMP.

---

[1] Visualization Toolkit file format (Kitware, Inc., 2006)

[2] Virtual Reality Modeling Language(International Organization for Standardization, 2009)

[3] Recursive acronym for *GNU's not Unix*

B.2 Development Tools

Following development tool have been used for this work: The

Eclipse platform in version 3.3.1.1 (The Eclipse Foundation, 2009) was applied as a development environment in combination with

CMake version 2.4.8 (Kitware, Inc., 2009a), a the cross-platform, open-source build system. An

Apache SVN adapter on Apache HTTP Server 2.2.6 (The Apache Software Foundation, 2009) was utilized with

SVN version 1.5 (CollabNet, 2009) to store the developed code decentralized. For MacOSX and Linux the

g++ compiler version 4.0.2, respectively 4.3 (The GCC steering committee, 2009) was used. The

MinGW version 3.14 (Msys 1.0.10, Gcc 3.4.5) was used (Peters *et al.*, 2009) for Windows.

B.3 Toolkits

A number of C++-toolkits have been used. For a cross-platform GUI[4] Trolltech's toolkit

Qt in version 4.3.3 (Qt Software, 2009) was used. To view three-dimensional scenarios, Kitware's

Visualization Toolkit (VTK) in version 5.0.4 (Schroeder *et al.*, 2006; Kitware, Inc., 2006, 2009d) was applied.

For image resampling filters and registration methods Kitware's

Insight Toolkit (ITK) in version 3.4.0 (Ibanez *et al.*, 2005; Yoo, 2004; Kitware, Inc., 2009c) was used.

QextSerialPort (Fosdick *et al.*, 2009) was utilized to encapsulate a serial port.

---

[4]Graphical User Interface

Appendix C

Voxel2MCNP Documents

C.1 Brief Instructions

In order to explain how to use Voxel2MCNP, a simple example is given and guides the user step by step through the procedure to finally generate an MCNP input file. The example here represents the generation of an ^{241}Am liver counting scenario in the PBC with the voxel model of the LLNL Torso Phantom. The handling is mostly intuitive. Most functions explain themselves by reading the menu items. So far the language of the software is German, but an English version is planned. A switch in the software is intended to switch between the languages and also between different number formats. Voxel2MCNP operates usually in the German mode, i.e. uses a decimal comma instead of a decimal point. This must be considered when working on international operating systems.

1. The installation of Voxel2MCNP on a computer is simple. For example, on Apple MacOSX, just get the latest `voxel2mcnp.app` file and copy it into your Applications folder. On Microsoft Windows, copy the latest `voxel2mcnp.exe` file in any new folder. Copy the belonging dynamic libraries (`.dll`) into the same folder. These are the Qt libraries with the file names Qt3Support4, QtOpenGL4, QtAssistantClient4, QtScript4, QtCore4, QtSql4, Qt-Cored4, QtSqld4, QtDesigner4, QtSvg4, QtDesignerComponents4, QtTest4, QtGui4, QtXml4, and QtNetwork4 and additionally the

MinGW dynamic library called mingwm10.

2. Open Voxel2MCNP.

3. Load a voxel model by choosing menu $\boxed{\text{Voxelmodell}}$ click action $\boxed{\text{Laden}}$. A file dialog pops up. Supported formats are kalattice (`.lat`), BMP-bitmap (file series), ASCII format, and 8-bit binary format. In this example the file

   `torso7_27stel_overlay4.lat`

   is chosen. It is the coarsened LLNL Torso Phantom model with overlay 4. The kalattice-format has already all information about spacing and dimensions of the voxel model. In case of other formats, a dialog pops up and asks for missing parameter. During the loading process, the 2D-viewer cycles through all slices of the voxel model. After loading the voxel model, the program window looks similar like the screenshot in figure 4.6 on page 49.

4. Choose an appropriate organ parameter set by choosing menu $\boxed{\text{Organe}}$ and click on $\boxed{\text{Torsophantom-Organparameter setzen}}$. This sets the organ parameter (the association of voxel ID# with density and material) coded in Voxel2MCNP. If necessary, the parameters can be modified by choosing $\boxed{\text{Organparameter editieren}}$. Here, the voxel ID# associated with the liver can be looked up (see figure C.1).

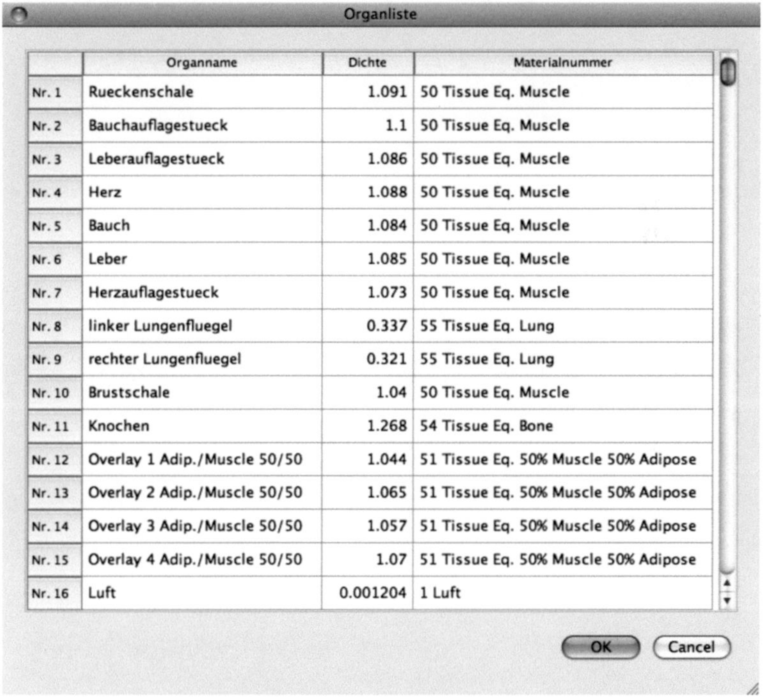

	Organname	Dichte	Materialnummer
Nr. 1	Rueckenschale	1.091	50 Tissue Eq. Muscle
Nr. 2	Bauchauflagestueck	1.1	50 Tissue Eq. Muscle
Nr. 3	Leberauflagestueck	1.086	50 Tissue Eq. Muscle
Nr. 4	Herz	1.088	50 Tissue Eq. Muscle
Nr. 5	Bauch	1.084	50 Tissue Eq. Muscle
Nr. 6	Leber	1.085	50 Tissue Eq. Muscle
Nr. 7	Herzauflagestueck	1.073	50 Tissue Eq. Muscle
Nr. 8	linker Lungenfluegel	0.337	55 Tissue Eq. Lung
Nr. 9	rechter Lungenfluegel	0.321	55 Tissue Eq. Lung
Nr. 10	Brustschale	1.04	50 Tissue Eq. Muscle
Nr. 11	Knochen	1.268	54 Tissue Eq. Bone
Nr. 12	Overlay 1 Adip./Muscle 50/50	1.044	51 Tissue Eq. 50% Muscle 50% Adipose
Nr. 13	Overlay 2 Adip./Muscle 50/50	1.065	51 Tissue Eq. 50% Muscle 50% Adipose
Nr. 14	Overlay 3 Adip./Muscle 50/50	1.057	51 Tissue Eq. 50% Muscle 50% Adipose
Nr. 15	Overlay 4 Adip./Muscle 50/50	1.07	51 Tissue Eq. 50% Muscle 50% Adipose
Nr. 16	Luft	0.001204	1 Luft

Figure C.1: Organ parameter list.

5. Edit Parameters by choosing in menu Parameter and hit Editieren (see figure C.2). The source organ (parameter #127) can be changed to liver (voxel ID# 6) by simply overwriting the default value. Parameter #126 sets the source nuclide, it is already set to one (^{241}Am), hence no action is needed. Many other parameters can be set. In this example the default values are kept.

Parameter	Wert	Einheit
mk[114] Luftraum in der Messkammer, z–Startpunkt	0	mm
mk[115] Luftraum in der Messkammer, z–Endpunkt	4196.2	mm
mk[116] Schichtdicke Kupfer	0.5	mm
mk[117] Schichtdicke Zinn	1.5	mm
mk[118] Schichtdicke Blei	5	mm
mk[119] Schichtdicke Stahl	150	mm
mk[120] Schalter Phoswich	1	1 = ein
mk[121] Schalter HPGe	0	1 = ein
mk[122] Schalter fuer den Elektronenmodus	1	1 = ein
mk[123] NPS / Histories	1e+07	–
mk[124] Schalter fuer das Lungenquaderexperiment	0	1 = ein
mk[125] Nummer der Photonen Cross Section Bibliothek	4	–
mk[126] Nuklidnummer, (1)=Am–241 (2)=Cs–137 (3)=Co...	1	–
mk[127] Quellorgannummer	17	–
mk[128] Quellvoxelanzahl, (0)=alle	0	–

Figure C.2: Parameter list.

6. Now, look at the 3D-visualization by choosing Visualisierung and hit vtkBetrachter. A new window pops up that looks similar to the one in figure 4.7. The window can be enlarged by clicking in the right-bottom corner and dragging the windows to the desired size. The combo boxes in upper-left corner let the user choose an organ of interest. After choosing, it appears in the VTK-viewer

area. Objects, for instance the examination table (German: *Liege*), appear and disappear by clicking on the check boxes on the left side. The opacity of the first five organs can be changes in the spin-box below. The viewing angle can be changed by clicking and dragging the mouse in various directions in the box of the VTK-viewer. Doing the same in combination with holding the *Shift*-key slides the view on a plane. Figure C.3 shows the stereo mode of the viewer, which gives the user a three-dimensional impression when wearing regular red-blue glasses. The mode can be activated and de-activated by hitting the *3*-key. Details of a visualization can be also seen in figure 6.27. It is recommended to check the position of the liver to be sure that the voxel model has the liver on the right side. If this is not the case the model can be mirrored (see point 8). The parameters of the phoswich rack (X_{PW}, W_{HA}, etc.) can be changed in the spin-boxes. The labels for each parameter are the same like described in section 6.3.1. The change is visualized instantly and can be checked in the viewer. When the detector position is in the desired position, the OK button can be clicked to transfer the phoswich parameters to the parameter list.

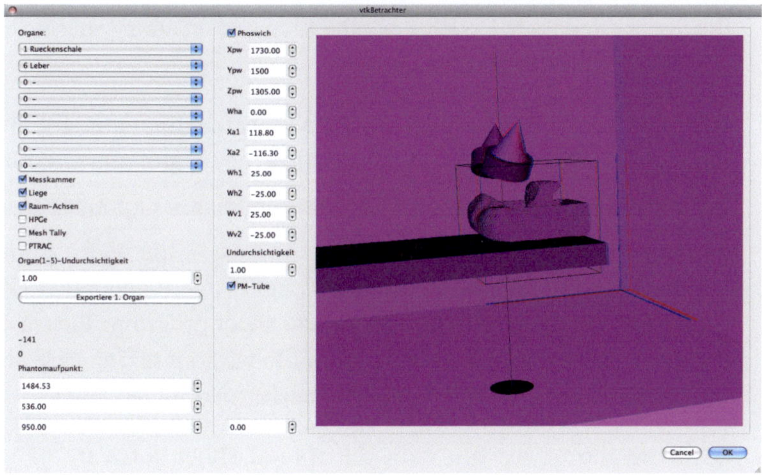

Figure C.3: Stereo visualization of the scenario. Using regular red-blue glasses provide a three-dimensional experience.

7. The parameter list can be saved and re-loaded later for other voxel models in the menu Parameter .

8. If the voxel model must be mirrored or rotated, the user can choose in the menu Voxelmodel actions like Spiegelung an x-y-Ebene (mirror on x-y-plane) or Modell um 90° um y-Achse drehen (rotate 90° around y-axis).

9. If all parameters are set to the desired values, the MCNP input file can be generated. This is done by clicking on Input file erzeugen in the menu MCNP . A file dialog lets the user choose a folder and file name. It is recommended to choose a file name with the ending `.i`, for instance `liver_scenario.i`. If needed, the generated input file can be edited manually with any ASCII-text editor.

10. If MCNPX is installed on the computer, the input file can be started with the shell command

    `mcnpx i=liver_scenario.i n=liver_scenario.`

    Note that using this way the n-option generates an output file with the full file name `liver_scenario.o`. Also other associated files generated by MCNPX have the filename `liver_scenario` with a specific file ending.

11. After the simulation is finished, the user can extract the tally information from the output file. The action csv-Datei aus Outputfile or, in case of more than one file, csv-Datei aus Outputfile Batch in the menu MCNP can be clicked to do this. A file dialog pops up and asks for the path of the file(s). Voxel2MCNP generates `.csv` file(s) that can be read with any spread sheet program. Please note that VoxelMCNP converts the output file without taking care about the results from the statistical tests performed by MCNPX.

12. Evaluate the `.csv` file with the spread sheet software of your choice.

C.2 Source Files

Table C.1: Voxel2MCNP's source code files and their tasks.

File	Task
`dasinfofenster.cpp`	information about Voxel2MCNP
`dimensionsabfrage.cpp`	GUI for voxel model dimensions
`faltung.cpp`	experimental spectrum folding feature
`federmodell.cpp`	experimental mass-spring model
`GageCheck.cpp`	interface to GageChek (section 6.3.1)
`hauptfenster.cpp`	GUI of main window
`igor.cpp`	generates WBC input files with IGOR
`itkTester.cpp`	interface to ITK features
`legende.cpp`	displays a voxel color legend
`lungcounter.cpp`	generates PBC voxel model input files
`lungcounter2.cpp`	generates PBC with XtRa detector
`main.cpp`	initializes the main window
`materialeditor.cpp`	manages materials
`messszenario.cpp`	VRML output of 3D-scenarios
`modifikationsart.cpp`	GUI for organ displacements
`nuklidbibliothek.cpp`	manages radionuclides
`organparameter.cpp`	manages organ parameters
`organsegmentabfrage.cpp`	GUI for adding organs
`organumnumerierung.cpp`	GUI for redefining organs
`output2csv.cpp`	extracts spectra from output files
`parameterliste.cpp`	manages parameter of scenario
`pfadesetzen.cpp`	GUI for setting paths
`populationmodell.cpp`	generates population models
`punktquelle.cpp`	generates PBC point source input files
`qextserialbase.cpp`	manages serial interface
`qextserialport.cpp`	manages RS-232 port
`regiongrowing.cpp`	region growing algorithm
`rollingballon.cpp`	experimental rolling ball algorithm
`schnittparameter.cpp`	GUI for cutting voxel model
`voxeldimensionsabfrage.cpp`	GUI for voxel dimensions
`voxelmodell.cpp`	generates VRML files from organs
`vrml_voxelmodelloptionen.cpp`	GUI for VRML file generation
`vtkBetrachter.cpp`	VTK-based 3D-viewer
`win_qextserialport.cpp`	manages RS-232 port on Windows

C.3   Example Input File

The following input file describes an ^{241}Am-point source on a steel table in the PBC, measured by two phoswich detectors with and without anti-coincidence feature. Note that MCNPX in version 2.6 should be used to run this input file, due to a bug in the anti-coincidence feature of version 2.5. Lines starting with c and all text following an $ are comments. Each entry in the input file is called *card*.

The first part describes the *cells* or the objects of the scenario. The cells are numbered here from 501 to 580. They are boolean combinations of geometric objects (starting at the 4th column). Each cell has a given material and density (second and third column).

A blank line separates the second part that describes the basic geometry of the cells. Each basic geometry is numbered from 2 to 85. Mostly so called *macrobody* cards are used. A rectangular box is coded by a rpp card, a cylinder is represented by rcc. The parameters are coordinates, distances and radii.

Another blank line separates the data card from the geometry cards. Here the mode is defined so that only electrons and photons are considered in the simulation. Material cards follow, which specify isotopic composition and the cross-section library that must be used. The SDEF card defines the source particle, its energy, and starting position. It is followed by the source energy spectrum. Finally the tallies of four phoswich detectors are defined with the corresponding cell number, the energy range, the bin width, and the parameters of the Gaussian broadening feature. In addition, the last four tallies have an anti-coincidence feature indicated by the phl card. The last two cards define the number of particle histories and the time period, after which MCNPX should save the data.

```
c Lungenzaehler MCNP Inputdatei erzeugt von
c VOXEL2MCNP
c (c)2006-2008 HS-KES/Institut fuer Strahlenforschung,
c FZK GmbH, KIT, http://www.hs-kes.de
c Erstellung: 14:17:06, 24.11.2008
    501      0              2 $ Void
    502     67   -7.87     -2  3 $ Stahlabschirmung
    503      4  -11.3      -3  4 $ Bleiabschirmung
    504      5   -7.31     -4  5 $ Zinnabschirmung
```

```
    505       6  -8.96        -5 6 $ Kupferabschirmung
c linker Phoswichdetektor
    510       9  -1.84       -12.3  -7.2 -11.1 $ Be-Eintrittsfenster
    511      10  -1.0          7.2  -8.2 -11.1 $ Zwischenschicht aus Al2O3
    512       7  -3.667        8.2  -9.2 -10.1 vol=32.429279 $ NaI(Tl)-Det.
    513       8  -4.51         9.2 -10.2 -10.1 vol=1647.4074 $ CsI-Det.
    514       7  -3.667       10.2 -11.2 -10.1 $ NaI-Kristall
    515      63  -7.87        11.2 -12.2 -12.1 $ Eisenscheibe
    516      10  -1.0          8.2 -11.2 -11.1 10.1 $ Al2O3
    517      63  -7.87       -12.3 -11.2 -12.1 11.1 $ Fe-Huelle
c rechter Phoswichdetektor
    520       9  -1.84       -18.3 -13.2 -17.1 $ Be-Eintrittsfenster
    521      10  -1.0         13.2 -14.2 -17.1 $ Zwischenschicht aus Al2O3
    522       7  -3.667       14.2 -15.2 -16.1 vol=32.429279 $ NaI(Tl)-Det.
    523       8  -4.51        15.2 -16.2 -16.1 vol=1647.4074 $ CsI-Det.
    524       7  -3.667       16.2 -17.2 -16.1 $ NaI-Kristall
    525      63  -7.87        17.2 -18.2 -18.1 $ Eisenscheibe
    526      10  -1.0         14.2 -17.2 -17.1 16.1 $ Al2O3
    527      63  -7.87       -18.3 -17.2 -18.1 17.1 $ Fe-Huelle
    560      72  -1.0         -60 $ Probandenliege - Kunstleder PVC
    561      74  -0.5         -61 $ Probandenliege - Schaumstoff PU
    562      74  -1.2         -62 $ Probandenliege - Hartplastik PU
    563      61  -7.87        -63 $ Probandenliege - Stahlrahmen
    570      73  -0.95        -70:-71:-72 $ Stahltisch - PE-Auflage
    571      61  -7.87        -73:-74:-75:-76:-77:-78:-79:-80:-81:-82:-83
    580      75  -1.05        -85 $ Quellfassung
    532       1  -0.00129      -6 12 18 60 61 62 63 &
70 71 72 73 74 75 76 77 78 79 80 81 82 83 85 $ Luft in der Messk.

c Geometriekarten
c Die verwendeten Parameter sind gespeichert unter
c <<W:/Allgemein/Intern/FWHM-Bestimmung Phoswich/
c Punktquellen-Untersuchung/nancys messungen/Messung 2/
c Messung2.csv>>
     2  rpp -15.7 240.52 &
-25.7 240.2 &
-15.7 435.32 $ aeussere Oberflaeche Stahlabschirmung
     3  rpp -0.7 225.52 &
-0.7 225.2 &
-0.7 420.32 $ Innere Oberflaeche Stahlabschirmung
     4  rpp -0.2 225.02 &
-0.2 224.7 &
-0.2 419.82 $ Innere Oberflaeche Bleiabschirmung
     5  rpp -0.05 224.87 &
-0.05 224.55 &
-0.05 419.67 $ Innere Oberflaeche Zinnabschirmung
     6  rpp 0 224.82 &
0 224.5 &
0 419.62 $ Innere Oberflaeche Kupferabschirmung
c Oberflaechen fuer den linken Phoswichdetektor:
     7  rcc 92.284412 98.128163 &
```

```
197.29044 0.0093456836 &
0.022749032 -0.0044877334 10.26
      8   rcc 92.284412 98.128163 &
197.29044 0.046728418 &
0.11374516 -0.022438667 10.16
      9   rcc 92.284412 98.128163 &
197.29044 0.084111153 &
0.20474129 -0.0403896 10.16
     10   rcc 92.284412 98.128163 &
197.29044 1.9831541 &
4.8273445 -0.95229702 10.16
     11   rcc 92.284412 98.128163 &
197.29044 2.9326755 &
7.1386462 -1.4082507 10.26
     12   rcc 92.284412 98.128163 &
197.29044 2.9700583 &
7.2296423 -1.4262017 10.29
c Oberflaechen fuer den rechten Phoswichdetektor:
     13   rcc 71.947876 98.140415 &
197.44837 -0.0095993741 &
0.0226761 -0.004318158 10.26
     14   rcc 71.947876 98.140415 &
197.44837 -0.047996871 &
0.1133805 -0.02159079 10.16
     15   rcc 71.947876 98.140415 &
197.44837 -0.086394367 &
0.2040849 -0.038863422 10.16
     16   rcc 71.947876 98.140415 &
197.44837 -2.0369872 &
4.8118685 -0.91631313 10.16
     17   rcc 71.947876 98.140415 &
197.44837 -3.0122836 &
7.1157603 -1.355038 10.26
     18   rcc 71.947876 98.140415 &
197.44837 -3.0506811 &
7.2064647 -1.3723106 10.29
     60   rpp 140 205 &
58.25 58.35 &
95 290 $ Probandenliege - Kunstleder
     61   rpp 140 205 &
54.25 58.25 &
95 290 $ Probandenliege - Schaumstoff
     62   rpp 140 205 &
49.25 54.25 &
95 290 $ Probandenliege - Hartplastik
     63   rpp 140 205 &
49.1 49.25 &
95 290 $ Probandenliege - Stahlrahmen
     70   rpp 57.54 107.04 &
77.105 77.18 &
163.67 229.67 $ Plastikauflage oben
```

```
   71   rpp 57.54 107.04 &
41.105 41.18 &
163.67 229.67 $ Plastikauflage mitte
   72   rpp 57.54 107.04 &
18.605 18.68 &
163.67 229.67 $ Plastikauflage unten
   73   rpp 57.54 107.04 &
76.865 77.105 &
163.67 229.67 $ Stahlplatte oben
   74   rpp 57.54 107.04 &
40.865 41.105 &
163.67 229.67 $ Stahlplatte mitte
   75   rpp 57.54 107.04 &
18.365 18.605 &
163.67 229.67 $ Stahlplatte unten
   76   rpp 57.26 57.54 &
0 77.18 &
163.39 168.19 $ Stahlteil West-Nord
   77   rpp 57.26 57.54 &
0 77.18 &
225.15 229.95 $ Stahlteil West-Sued
   78   rpp 107.04 107.32 &
0 77.18 &
163.39 168.19 $ Stahlteil Ost-Nord
   79   rpp 107.04 107.32 &
0 77.18 &
225.15 229.95 $ Stahlteil Ost-Sued
   80   rpp 57.54 62.06 &
0 77.18 &
163.39 163.67 $ Stahlteil Nord-West
   81   rpp 102.52 107.04 &
0 77.18 &
163.39 163.67 $ Stahlteil Nord-Ost
   82   rpp 57.54 62.06 &
0 77.18 &
229.67 229.95 $ Stahlteil Sued-West
   83   rpp 102.52 107.04 &
0 77.18 &
229.67 229.95 $ Stahlteil Sued-Ost
   85   rpp 81.49 82.59 &
77.18 77.38 &
195.455 197.805 $ Quellfassung

mode  p e
c Die verwendeten Materialien sind gespeichert unter
c  <<Default>>
m1 ELIB=03e PLIB=04p  $ Luft
      7000 -0.75518
      8000 -0.23135
     18000 -0.01288
m4 ELIB=03e PLIB=04p  $ Blei
```

```
        82000 1
m5  ELIB=03e PLIB=04p  $ Zinn
        50000 1
m6  ELIB=03e PLIB=04p  $ Kupfer
        29000 1
m7  ELIB=03e PLIB=04p  $ Natrium-Iodid
        11000 0.5
        53000 0.5
m8  ELIB=03e PLIB=04p  $ Caesium-Iodid
        53000 0.5
        55000 0.5
m9  ELIB=03e PLIB=04p  $ Beryllium
        4000 1
m10 ELIB=03e PLIB=04p  $ Aluminiumoxid Al2O3
        13000 2
        8000 3
m61 ELIB=03e PLIB=04p  $ Metall der Liege LC
        25000 -0.5
        26000 -99
m63 ELIB=03e PLIB=04p  $ Stahlgehaeuse Detektor LC
        24000 -0.7
        28000 -51.1
        29000 -8.2
        26000 -38.1
        73000 -1.2
m67 ELIB=03e PLIB=04p  $ Messkammerabschirmung aussen LC
        25000 -0.8
        26000 -98.7
m72 ELIB=03e PLIB=04p  $ Polyvinylchlorid
        1000 3
        6000 2
        17000 1
m73 ELIB=03e PLIB=04p  $ Polyethylen
        1000 2
        6000 1
m74 ELIB=03e PLIB=04p  $ Polyurethan
        1000 12
        6000 11
        7000 2
        8000 4
m75 ELIB=03e PLIB=04p  $ Acrylonitrile-Butadiene-Styrene
        1000 17
        6000 15
        7000 1
imp:p 0 1 27r
imp:e 0 1 27r
SDEF SUR=0 ERG=D2 PAR=2 POS= 82.04 77.28 196.63
c Als Strahlenquelle wurde Nuklid 1 (Am-241) gewaehlt.
c Hinweis: (1)=Am-241 (2)=Cs-137 (3)=Co-60 (4)=K-40
c (5)=Pu-238 (6)=Ba-133
c (7)=Na-22 (8)=Pb-210 (9)=Pu-239 (10)=Co57
```

```
c Halbwertszeit (in Sekunden): 1.36382e+10
c alle Angaben nach
c PTB-Bericht Schoetzig/Schrader, Mai 2000
#     si2    sp2
      L       D
      0.01187   0.00837
      0.01376   0.0108
      0.01395   0.1193
      0.01586   0.00377
      0.01754   0.1861
      0.02101   0.0482
      0.02634   0.024
      0.03218   0.000174
      0.0332    0.00126
      0.04273   5.5e-05
      0.04342   0.00073
      0.05556   0.000181
      0.05785   5.2e-05
      0.05954   0.359
      0.06483   1.45e-06
      0.06745   4.2e-06
      0.06976   2.9e-05
      0.0758    5.9e-06
      0.09897   0.000203
      0.10298   0.000195
      0.12301   1e-05
      0.1253    4.08e-05
      0.14655   4.61e-06
      0.16956   1.73e-06
      0.20801   7.91e-06
      0.32252   1.52e-06
      0.33235   1.49e-06
      0.33537   4.96e-06
      0.36734   7.9e-06
      0.36865   2.17e-06
      0.37665   1.383e-06
      0.6624    3.64e-06
      0.72201   1.96e-06
c ----normal-tallies--------------------------------
f8:p,e 512
fc8 NaI-Spektrum links, Det. 5
ft8 GEB -0.013 0.14 0
e8 0 1E-5 0.00097656 254I 0.25
f18:p,e 513
fc18 CsI-Spektrum links, Det. 6
ft18 GEB -0.038 0.16 0
e18 0 1E-5 0.0117188 254I 3
f28:p,e 522
fc28 NaI-Spektrum rechts, Det. 7
ft28 GEB -0.009 0.099 0
e28 0 1E-5 0.00097656 254I 0.25
```

```
f38:p,e 523
fc38 CsI-Spektrum rechts, Det. 8
ft38 GEB -0.017 0.12 0
e38 0 1E-5 0.0117188 254I 3
c ----anticoincidence-tallies----------------------------------
fc48 NaI-Spektrum, Det. 5 (anticoincidence)
f6:e  512  $ NaI
f16:e 513  $ CsI
ft6  GEB -0.013 0.14 0
ft16 GEB -0.038 0.16 0
f48:p,e 512   $ NaI
ft48  phl 1 6 1 1 16 1
e48 0 1E-5 0.00097656 254I 0.25
fu48 0.005 $ treshold value
fq48 e u
fc68 NaI-Spektrum, Det. 7 (anticoincidence)
f26:e 522  $ NaI
f36:e 523  $ CsI
ft26 GEB -0.009 0.099 0
ft36 GEB -0.017 0.12 0
f68:p,e 522   $ NaI
ft68  phl 1 26 1 1 36 1
e68  0 1E-5 0.00097656 254I 0.25
fu68 0.005 $ treshold value
fq68 e u
fc58 CsI-Spektrum, Det. 6 (anticoincidence)
c corresponding f6 tallies defined earlier
f58:p,e 513   $ CsI
ft58  phl 1 16 1 1 6 1
e58 0 1E-5 0.0117188 254I 3
fu58 0.005 $ treshold value
fq58 e u
fc78 CsI-Spektrum, Det. 8 (anticoincidence)
c corresponding f6 tallies defined earlier
f78:p,e 523   $ CsI
ft78  phl 1 36 1 1 26 1
e78 0 1E-5 0.0117188 254I 3
fu78 0.005 $ treshold value
fq78 e u
nps 10000000
prdmp j -120 1 3 1000000
```

Bibliography

Ackerman, M. J. The visible human project. *J Biocommun*, 18(2):14, 1991.

Agostinelli, S., Allison, J., Amako, K., Apostolakis, J., Araujo, H., Arce, P., Asai, M., Axen, D., *et al.* G4 - a simulation toolkit. *Nuclear Instruments and Methods in Physics Research A*, 506(3):250–303, 2003.

Akkurt, H. and Eckerman, K. F. Development of pimal: Mathematical phantom with moving arms and legs. Technical Report ORNL/TM-2007/14, Oak Ridge National Laboratory, 2007.

Allison, J., Amako, K., Apostolakis, J., Araujo, H., Dubois, P., Asai, M., Barrand, G., Capra, R., *et al.* Geant4 developments and applications. *IEEE Transactions on Nuclear Science*, 53(1):270–278, 2006.

Anderson, A. L., Campbell, G. W., and Griffith, R. V. In vivo measurement of acinides in the human lung. Technical Report UCRL 83548, 103-114, Lawrence Livermore National Laboratories, 1979.

Apple, Inc. MacOSX Leopard. 2009. URL `http://www.apple.com/macosx/` last checked in January 2009.

Argonne National Laboratory. MPI-2: Extensions to the Message-Passing Interface. 2009a. URL `http://www.mpi-forum.org/docs/mpi-20-html/mpi2-report.html` last checked in January 2009.

Argonne National Laboratory. MPICH-A Portable Implementation of MPI. 2009b. URL `http://www-unix.mcs.anl.gov/mpi/mpich1/` last checked in January 2009.

Armato, S., Giger, M., Moran, C., Blackburn, J., Doi, K., and MacMahon, H. Computerized detection of pulmonary nodules on CT scans. *Radiographics*, 19:1303–1311, 1999.

Baccaro, S., Barone, L. M., Borgia, B., Castelli, F., Cavallari, F., Dafinei, I., de Notaristefani, F., Diemoz, M., *et al.* Precise determination of the light yield of scintillating crystals. *Nuclear Instruments and Methods in Physics Research A*, 385:69–73, 1997.

Berger, C. D. and Lane, B. H. Biometric estimation of chest wall thickness of females. *Health Phys*, 49(3):419–424, 1985.

Besl, P. J. and McKay, N. D. A Method for Registration of 3-D Shapes. *IEEE Transactions on Pattern Analysis and Machine Intelligence*, 14(2):239–256, 1992.

Breustedt, B. *Entwicklung und Implementation von Monte-Carlo-Simulationen zur Auswertung von Messungen mit dem Kölner Ganzkörperzähler.* Ph.D. thesis, Universität Köln, 2004.

BUM. Richtlinie für die physikalische Strahlenschutzkontrolle zur Ermittlung der Körperdosis, Anlage zum Rundschreiben, Teil 2. Technical report, Bundesministerium für Umwelt, Naturschutz und Reaktorsicherheit, 2007.

Canberra Industries, Inc. *Genie 2000 Spectroscopy Software Customization Tools v 3.1*, 2006.

Canberra Industries, Inc. Genie^{TM} 2000 Gamma Analysis Software. 2009. URL `http://www.canberra.com/products/835.asp` last checked in January 2009.

Caon, M. Voxel-based computational models of real human anatomy: a review. *Radiat Environ Biophys*, 42(4):229–235, 2004.

CEN. Size designation of clothes - Part 1: Terms, definitions and body measurement procedure. Norm EN 13402-1:2001 (ISO 3635:1981 modified), European Committee for Standardization, 2001.

CEN. Size designation of clothes - Part 2: Primary and secondary dimensions. Norm EN 13402-2:2002, European Committee for Standardization, 2002.

CEN. Size designation of clothes - Part 3: Measurements and intervals. Norm EN 13402-3:2004, European Committee for Standardization, 2004.

CollabNet. Subversion, an open source version control system. 2009. URL `http://subversion.tigris.org/` last checked in January 2009.

Cordes, G. Effizienzkalibrierung des Teilkörperzählers mit Phoswich-Detektoren. Standard Operating Procedure SAA-IVM-203 Rev. 1, Hauptabteilung Sicherheit - KES, 2007.

Cristy, M. Mathematical phantoms representing children of various ages for use in estimates of internal dose. Technical Report NUREG/CR-1159, Oak Ridge National Laboratory, 1980.

Dean, P. N. Estimation of chest wall thickness in lung counting for plutonium. *Health Phys*, 24(4):439–441, 1973.

Debertin, K. and Helmer, R. G. *Gamma- and X-ray spectrometry with semiconductors*. North-Holland, 1988.

Dimbylow, P. J., editor. *Voxel phantom development: proc. int. workshop*. National Radiological Protection Board, Chilton, UK, 1996.

Doerfel, H., Heide, B., and Sohlin, M. Entwicklung eines Verfahrens zur numerischen Kalibrierung von Teilkörperzählern. Wissenschaftlicher Bericht FZKA 7238, Forschungszentrum Karlsruhe, 2006.

El Bakkari, B. The Monte Carlo simulation of a Package formed by the combination of three scintillators: Brillance380, Brillance350, and

Prelude420. Internal Report I3 506065 JRA9 – RHIB, EURONS, 2006.

Fosdick, B., Sander, S., and Policht, M. Qextserialport, a cross-platform serial port class. 2009. URL `http://qextserialport.sourceforge.net/` last checked in January 2009.

Fry, F. A. and Sumerling, T. Measurement of chest wall thickness for assessment of plutonium in human lungs. *Health Phys*, 39(1):89–92, 1980.

Gómez-Ros, J. M., de Carlan, L., Franck, D., Gualdrini, G., Lis, M., López, M. A., Moraleda, M., Zankl, M., *et al.* Monte carlo modelling of germanium detectors for the measurement of low energy photons in internal dosimetry: Results of an international comparison. *Radiation Measurements*, 43:510–515, 2008.

Griffith, R., Dean, P. N., Anderson, A. L., and Fisher, J. C. Fabrication of a tissue-equivalent torso phantom for intercalibration of in-vivo transuranic-nuclide couning facilities. In *Symp. on Advances in Radiation Protection Monitoring*. Stockholm, Sweden, 1978.

Griffith, R. V. Polyurethane as a base for a famliy of tissue equivalent materials. In *Proc. 5th International Congress, IRPA, International Meeting Commission*. IRPA, Jerusalem, Israel, 1980.

Griffith, R. V., Anderson, A. L., Dean, P. N., Fisher, J. C., and Sundbeck, C. W. Tissue-equivalent torso phantom for calibration of transuranic-nuclide counting facilities. In *Radiobioassay and internal dosimetry workshop, Albuquerque, NM*. 1986.

Gropp, W., Lusk, E., Doss, N., and Skjellum, A. A high-performance, portable implementation of the MPI message passing interface standard. *Parallel Computing*, 22(6):789–828, 1996.

GSF. Berechnung von Organdosen und Dosisverteilungen bei externer und interner Exposition mit Hilfe von Strahlentransportrechnungen in Voxelmodellen. Schriftenreihe BMU-2006-691, BMU, Bonn, Germany, 2006.

Halbleib, J. A., Kensek, R. P., Valdez, G. D., Mehldorn, T. A., Seltzer, S. M., and Berger, M. J. ITS Version 3.0: The Integrated TIGER Series of Coupled Electron/Photon Monte Carlo Codes. *IEEE Transactions on Nuclear Science*, 39:1025–1030, 1992.

Hegenbart, L. Phoswich-Positionserfassungssystem in der Teilkörperzählkammer. Laborinterner Bericht, Forschungszentrum Karlsruhe, Hauptabteilung Sicherheit - KES, 2008.

Hegenbart, L., Marzocchi, O., Breustedt, B., and Urban, M. Validation of a Monte Carlo efficiency calibration procedure for a partial body counter system with a voxel model of the LLNL Torso Phantom. *Radiation Protection Dosimetry*, accepted 19-Feb, 2009.

Hegenbart, L., Na, Y. H., Zhang, J. Y., Urban, M., and Xu, X. G. A Monte Carlo study of lung counting efficiency for female workers of different breast sizes using deformable phantoms. *Phys Med Biol*, 53(19):5527–5538, 2008.

Hendricks, J. S., McKinney, G. W., Fensin, M. L., James, M. R., Johns, R. C., Durkee, J. W., Finch, J. P., Pelowitz, D. B., *et al.* MCNPX 2.6.0 Extensions. Technical Report LA-UR-08-2216, Los Alamos National Laboratory, 2008.

Hickman, D. P. LLNL Torso Phantom assembly and disassembly. Technical Report UCRL-TR-215939, Lawrence Livermore National Laboratories, 2005.

Hughes, H. G. Uncertainties beyond statistics in monte carlo simulations. *Radiation Protection Dosimetry*, 126(1-4):45–51, 2007.

IAEA. Direct methods for measuring radionuclides in the human body. Safety Series No. 114, IAEA, Vienna, Austria, 1996.

Ibanez, L., Schroeder, W., Ng, L., and Cates, J. *The ITK Software Guide*. Insight Software Consortium, 2nd edition, 2005. Updated for ITK version 2.4. URL `http://www.itk.org/ItkSoftwareGuide.pdf` last checked in January 2009.

IBT Universität Karlsruhe. kaLattice Sources Documentation. 2009. URL `http://www.ibt.uni-karlsruhe.de/software/kaLattice/doxygen/html/index.html` last checked in January 2009.

ICRP. *Report of the task group on reference man.* Number 23 in ICRP Publication. ICRP, 1975.

ICRP. *Basic anatomical and physiological data for use in radiological protection: reference values.* Number 89 in ICRP Publication. ICRP, 2002.

ICRU. *Phantoms and computational models in therapy, diagnosis and protection.* Number 48 in ICRU Report. ICRU, Bethesda, MD, 1992.

International Organization for Standardization. The Virtual Reality Modeling Language. 2009. URL `http://www.web3d.org/x3d/specifications/vrml/ISO-IEC-14772-VRML97/` last checked in January 2009.

Jamal, N., Ng, K. H., McLean, D., Looi, L. M., and Moosa, F. Mammographic breast glandularity in malaysian women: data derived from radiography. *AJR Am J Roentgenol*, 182(3):713–717, 2004.

Jarrousse, O., Fritz, T., and Dössel, O. A modified mass-spring system for myocardial mechanics modeling. In *4th European Conference of the International Federation for Medical and Biological Engineering*, volume 22 of *IFMBE Proceedings*. ECIFMBE, Antwerp, Belgium, 2008.

Jones, D. G. Use of voxel phantoms in organ dose calculations. In P. J. Dimbylow, editor, *Voxel phantom development: proc. an international workshop*. National Radiological Protection Board, Chilton, UK, 1996.

Kawrakow, I. Accurate condensed history Monte Carlo simulation of electron transport. I. EGSnrc, the new EGS4 version. *Med Phys*, 27(3):485–498, 2000.

Kitware, Inc. *The Visualization Toolkit User's Guide.* Kitware, 5th edition, 2006.

Kitware, Inc. CMake, the cross-platform, open-source build system. 2009a. URL http://www.cmake.org/ last checked in January 2009.

Kitware, Inc. Paraview, an open-source, multi-platform data analysis and visualization application. 2009b. URL http://www.paraview.org/ last checked in January 2009.

Kitware, Inc. The Insight Toolkit. 2009c. URL http://www.itk.org/ last checked in January 2009.

Kitware, Inc. The Visualisation Toolkit. 2009d. URL http://www.vtk.org/ last checked in January 2009.

Kovtun, A. N., Firsanov, V. B., Fominykh, V. I., and Isaakyan, G. A. Metrological parameters of the unified calibration whole-body phantom with gamma-emitting radionuclides. *Radiation Protection Dosimetry*, 89(3-4):239–242, 2000.

Kramer, G. HML's methods for lung counting for women, 2008. Email.

Kramer, G. H. and Capello, K. Effect of lung volume on counting efficiency: a monte carlo investigation. *Health Phys*, 88(4):357–363, 2005.

Kramer, G. H., Crowley, P., and Burns, L. C. The uncertainty in the activity estimate from a lung count due to the variability in chest wall thickness profile. *Health Phys*, 78(6):739–743, 2000a.

Kramer, G. H. and Hauck, B. M. The effect of lung deposition patterns on the activity estimate obtained from a large area germanium detector lung counter. *Health Phys*, 77(1):24–32, 1999.

Kramer, G. H., Lopez, M. A., and Webb, J. A joint HML-CIEMAT-CEMRC project: testing a function to fit counting efficiency of a lung counting germanium detector array to muscle-equivalent chest wall thickness and photon energy using a realistic torso phantom over an extended energy range. *Radiation Protection Dosimetry*, 92(4):324–327, 2000b.

Kramer, R., Zankl, M., Williams, G., and Drexler, G. The calculation of dose from external photon exposures using human phantoms and Monte Carlo methods. Part I: The male (Adam) and female (Eva) adult mathematical phantoms. Technical Report GSF-Bericht S-885, GSF-Forschungszentrum für Umwelt und Gesundheit, Neuherberg, Germany, 1982.

LANL. 50th anniversary article: Evolving from calculators to computers. 2009. URL `http://www.lanl.gov/history/atomicbomb/computers.shtml` last checked in January 2009.

Lee, C., Lee, C., Lodwick, D., and Bolch, W. E. NURBS-based 3-D anthropomorphic computational phantoms for radiation dosimetry applications. *Radiation Protection Dosimetry*, 127(1-4):227–232, 2007a.

Lee, C., Lee, C., Williams, J. L., and Bolch, W. E. Whole-body voxel phantoms of paediatric patients–UF Series B. *Phys Med Biol*, 51(18):4649–4661, 2006.

Lee, C., Lodwick, D., Hasenauer, D., Williams, J. L., Lee, C., and Bolch, W. E. Hybrid computational phantoms of the male and female newborn patient: Nurbs-based whole-body models. *Phys Med Biol*, 52(12):3309–3333, 2007b.

Lee, J. M., Appugliese, D., Kaciroti, N., Corwyn, R. F., Bradley, R. H., and Lumeng, J. C. Weight status in young girls and the onset of puberty. *Pediatrics*, 119(3):e624–30, 2007c.

Leone, D. and Hegenbart, L. Validation of Monte Carlo simulations with phoswich detectors and point sources. Annual Report 2008 FZKA 7476, Central Safety Department, FZK, Karlsruhe, Germany, 2009.

Lietzau, C. Anatomium - P1 dataset. 2009. URL `http://www.anatomium.com/` last checked in January 2009.

Liye, L., Franck, D., de Carlan, L., and Junli, L. Application of Monte Carlo calculation and OEDIPE software for virtual calibration of an in vivo counting system. *Radiation Protection Dosimetry*, 127(1-4):282–286, 2007.

Loughlin, M., Wareing, T., Barnett, A., Failla, G., and McGhee, J. Comparison of Attila and MCNP for Fusion Applications. In *M&C 2005: International Topical Meeting on Mathematics and Computation, Supercomputing, Reactor Physics, and Nuclear and Biological Applications*. Avignon, France, 2005.

Marzocchi, O. Characterisation of a Canberra Cryo Pulse CP-5 Radiation Detector. Annual Report 2007 FZKA 7410, Central Safety Department, FZK, Karlsruhe, Germany, 2008.

McKinney, G. W. Differences MCNPX 2.5.0 and 2.6.0 about PHL, 2008. Email.

Merck. Aluminum Oxide, Safety Data Sheet. 2009. URL `http://www.merck-chemicals.com/documents/sds/emd/deu/de/1010/101095.pdf` last checked in January 2009.

Microsoft Corporation. Microsoft Excel for Mac. 2009a. URL `http://www.microsoft.com/mac/products/excel2008/default.mspx#/building_charts/` last checked in January 2009.

Microsoft Corporation. Windows XP. 2009b. URL `http://www.microsoft.com/germany/windows/windows-xp/` last checked in January 2009.

Mohr, U. and Breustedt, B. Messung von inkorporierten Radionukliden mittels Gammaspektrometrie im Teilkörperzähler mit Phoswichdetektor. Method Description MB-HS-012 Rev. 1, Hauptabteilung Sicherheit - KES, 2007.

Mohr, U. and Cordes, C. Qualitätssicherung am Teilkörperzähler mit Phoswich-Detektoren. Standard Operating Procedure SAA-IVM-200 Rev. 3, Hauptabteilung Sicherheit - KES, 2008.

Na, Y. H., Zhang, J. Y., and Xu, X. G. A mesh-based anatomical deformation method for creating size-adjustable whole-body patient models. In *AAPM 50th Annual Meeting*. Houston, Texas, 2008.

Nelson, W. R., Hirayama, H., and Rogers, D. W. O. The EGS4 Code System. Technical Report SLAC-265, Stanford Linear Accelerator, 1985.

NEMA. DICOM - strategic document. 2008. URL `http://medical.nema.org/dicom/geninfo/Strategy.pdf` last checked in January 2009.

Nooruddin, F. and Turk, G. Simplification and repair of polygonal models using volumetric techniques. *IEEE Trans. Vis. Comput. Graphics*, 9:191–205, 2003.

Odgen, C. L., Carroll, M. D., Curtin, L. R., McDowell, M. A., Tabak, C. J., and Flegal, K. M. Prevalence of overweight and obesity in the United States, 1999-2004. *JAMA*, 295:1549–1555, 2006.

Ortwin Rave. MolyCom - Technische Daten. 2009. URL `http://www.rave-minerals.com/fertiggranulate.HTM` last checked in January 2009.

Palmer, H. E., Rieksts, G. A., Jefferies, S. J., and Gunston, K. J. Improved counting efficiencies for measuring 239pu in the lung in the sitting position. *Health Phys*, 57(5):747–752, 1989.

Pelowitz, D. G. MCNPX User's Manual version 2.5.0. Technical Report LA-CP-05–0369, Los Alamos National Laboratory, 2005.

Peters *et al.* Minimalist GNU for Windows. 2009. URL `http://www.mingw.org/` last checked in January 2009.

Petoussi-Henss, N., Zankl, M., Fill, U., and Regulla, D. The GSF family of voxel phantoms. *Phys Med Biol*, 47(1):89–106, 2002.

Qt Software. Qt Cross-Platform Application Framework. 2009. URL `http://www.qtsoftware.com/products` last checked in January 2009.

Radiology Support Devices. *LLNL Realistic Phantom - Technical Data and Instruction Manual*, 1985.

Research and Technical Centre "Protection". Technical documents for human whole body phantom. Technical report, Research and Technical Centre "Protection", Saint Petersburg, Russia, 1997.

Rosset, A., Spadola, L., and Ratib, O. Osirix: an open-source software for navigating in multidimensional dicom images. *J Digit Imaging*, 17(3):205–216, 2004.

Rueckert, D., Sonoda, L. I., Hayes, C., Hill, D. L., Leach, M. O., and Hawkes, D. J. Nonrigid registration using free-form deformations: application to breast mr images. *IEEE Trans Med Imaging*, 18(8):712–721, 1999.

Sachse, F. B., Werner, C. D., Meyer-Waarden, K., and Dössel, O. Development of a human body model for numerical calculation of electrical fields. *Comput Med Imaging Graph*, 24(3):165–171, 2000.

Salvat, F., Fernandez-Varea, J. M., and Sempau, J. PENELOPE-2006, A Code System for Monte Carlo Simulation of Electron and Photon Transport. Technical Report NEA-6222, Barcelona, Spain, 2006.

Schläger, M. Precise modelling of coaxial germanium detectors in preparation for a mathematical calibration. *Nuclear Instruments and Methods in Physics Research A*, 580:137–140, 2007.

Schötzig, U. and Schrader, H. *Halbwertszeiten und Photonen-Emissionswahrscheinlichkeiten von häufig verwendeten Radionukliden*. PTB-RA16/5. PTB, Braunschweig, 5th edition, 2000.

Schroeder, W., Martin, K., and Lorensen, B. *The Visualization Toolkit, An Object-Oriented Approach To 3D Graphics*. Kitware, 2006.

Sessler, S. Numerische Simulation von im Karlsruher Ganzkörperzähler gemessenen IGOR-Spektren. Diplomarbeit, 2007.

Snyder, W. S., Fisher, H. L. J., Ford, M. R., and Warner, G. G. Estimates of absorbed fractions for monoenergetic photon sources uniformly distributed in various organs of a heterogeneous phantom. *J Nucl Med*, 3:7–52, 1969.

Sohlin, M. *Implementation of an Anthropomorphous Voxel Phantom in the MCNP5 Code.* Master's thesis, Göteborg University, 2005.

Taranenko, V., Zankl, M., and Schlattl, H. Voxel phantom setup in mcnpx. In *The Monte Carlo Method: Versatility Unbounded In A Dynamic Computing World.* American Nuclear Society, 2005.

Taylor, F. Y. *History of the Lawrence Livermore National Laboratory Torso Phantom.* Master's thesis, San Jose State University, 1997.

The Apache Software Foundation. HTTP Server Project. 2009. URL `http://httpd.apache.org/` last checked in January 2009.

The C++ Standards Committee. JTC1/SC22/WG21 - The C++ Standards Committee. 2009. URL `http://www.open-std.org/jtc1/sc22/wg21/` last checked in January 2009.

The Consortium of Computational Human Phantoms. The Consortium of Computational Human Phantoms. 2009. URL `http://www.virtualphantoms.org/phantoms.htm` last checked in January 2009.

The Eclipse Foundation. Eclipse, an open source development platform. 2009. URL `http://www.eclipse.org/` last checked in January 2009.

The GCC steering committee. GCC, the GNU Compiler Collection. 2009. URL `http://gcc.gnu.org/` last checked in January 2009.

The GIMP Team. GIMP, the GNU Image Manipulation Program. 2009. URL `http://www.gimp.org/` last checked in January 2009.

The OsiriX Foundation. OsiriX, an advanced Open-Source PACS Workstation DICOM viewer. 2009. URL `http://www.osirix-viewer.com/` last checked in January 2009.

The Ubuntu Community. Ubuntu. 2009. URL `http://www.ubuntu.com/` last checked in January 2009.

van Heesch, D. Doxygen documentation system. 2009. URL `http://www.stack.nl/~dimitri/doxygen/` last checked in January 2009.

van Riper, K. A. Bodybuilder. 2009. URL `http://www.whiterockscience.com/bodybuilder/bodybuilder.html` last checked in January 2009.

Vollmer, J., Mencl, R., and Müller, H. Improved Laplacian Smoothing of Noisy Surface Meshes. *Computer Graphics Forum*, 18:131–138, 1999.

Wang, B., Xu, X. G., Goorley, J. T., and Bozkurt, A. Issues related to the use of mcnp code for an extremely large voxel model vip-man. In *The Monte Carlo Method: Versatility Unbounded In A Dynamic Computing World*. American Nuclear Society, 2005.

Williams, T., Kelley, C., Lang, R., Kotz, D., Campbell, J., Elber, G., and Woo, A. Gnuplot, a portable command-line driven plotting utility. 2009. URL `http://www.gnuplot.info/` last checked in January 2009.

Willmann, S., Hohn, K., Edginton, A., Sevestre, M., Solodenko, J., Weiss, W., Lippert, J., and Schmitt, W. Development of a physiology-based whole-body population model for assessing the influence of individual variability on the pharmacokinetics of drugs. *J Pharmacokinet Pharmacodyn*, 34(3):401–431, 2007.

Xu, X. G., Chao, T. C., and Bozkurt, A. Vip-man: an image-based whole-body adult male model constructed from color photographs of the visible human project for multi-particle monte carlo calculations. *Health Phys*, 78(5):476–486, 2000.

Xu, X. G., Taranenko, V., Zhang, J. Y., and Shi, C. A boundary-representation method for designing whole-body radiation dosimetry models: pregnant females at the ends of three gestational periods–rpi-p3, -p6 and -p9. *Phys Med Biol*, 52(23):7023–7044, 2007.

Xu, X. G., Zhang, J. Y., and Na, Y. H. Preliminary data for mesh-based deformable phantom development: Is it possible to design person-specific phantoms on-demand. In *The International Conference on Radiation Shielding-11*. Pine Mountain, GA, 2008.

Yoo, T. S. *Insight into Images: Principles and Practice for Segmentation, Registration, and Image Analysis*. AK Peters, 1st edition, 2004.

Zankl, M., Becker, J., Fill, U., Petoussi-Henss, N., and Eckerman, K. F. GSF male and female adult voxel models representing ICRP Reference Man—the present status. In *The Monte Carlo Method: Versatility Unbounded In A Dynamic Computing World*. American Nuclear Society, 2005.

Zankl, M., Eckerman, K. F., and Bolch, W. E. Voxel-based models representing the male and female icrp reference adult–the skeleton. *Radiation Protection Dosimetry*, 127(1-4):174–186, 2007.

Zhang, B., Ma, J., Liu, L., and Cheng, J. CNMAN: a Chinese adult male voxel phantom constructed from color photographs of a visible anatomical data set. *Radiation Protection Dosimetry*, 124(2):130–136, 2007.

Zhang, J. Y., Xu, X. G., and Shi, C. Y. Four-dimensional human modeling for respiratory motion and the impact on radiation treatment planning. *Chin J Med Imaging Technol*, 22(9):1301–1305, 2006.

Zubal, I. G. The Zubal Phantoms. 2009. URL `http://noodle.med.yale.edu/zubal/` last checked in January 2009.

Zubal, I. G., Harrell, C. R., Smith, E. O., Rattner, Z., Gindi, G., and Hoffer, P. B. Computerized three-dimensional segmented human anatomy. *Med Phys*, 21(2):299–302, 1994.

List of Abbreviations

AC	Anti-coincidence
ASCII	American Standard Code for Information Interchange
BfS	Bundesamt für Strahlenschutz
BG	Bust girth
BMU	Bundesministerium für Umwelt, Naturschutz und Reaktorsicherheit
BREP	Boundary Representation
CAE	Computer aided engineering
CERN	Conseil Européen pour la Recherche Nucléaire
CIRP	China Institute for Radiation Protection
CS	Cup size
CWT	Chest wall thickness
EGS	Electron Gamma Shower - a Monte Carlo radiation transport code
EGSnrc	A version of EGS maintained at National Research Council of Canada
ENDF	Evaluated Nuclear Data File

EURADOS	European Radiation Dosimetry Group
FWHM	Full width at half maximum
Geant4	Geometry and Tracking 4 - a Monte Carlo radiation transport code
GSF	Gesellschaft für Strahlenforschung
GUI	Graphical User Interface
HML	Human Monitoring Laboratory
IBT	Institut für Biomedizinische Technik
ICP	Iterative Closest Point
ICRP	International Commission on Radiological Protection
ICRU	International Commission on Radiation Units and Measurements
ISF	Institut für Strahlenforschung
ITK	Insight Toolkit
IVM	In vivo Messlabor
KIT	Karlsruhe Institute of Technology
LLNL	Lawrence Livermore National Laboratory
MCA	Multi channel analyzer
MCNP	Monte Carlo N-Particle
MCNPX	Monte Carlo N-Particle eXtended
MEET Man	Models for simulation of Electromagnetic, Elastomechanic, and Thermic Behaviour of Man
MPI	Message Passing Interface

MRI	Magnetic Resonance Imaging
NNDC	National Nuclear Data Center
NURBS	Non-uniform rational B-splines
ORNL	Oak Ridge National Laboratory
PBC	Partial Body Counter
PENELOPE	Penetration and Energy Loss of Positrons and Electrons - a Monte Carlo radiation transport code
PET	Positron Emission Tomography
PMB	Physics in Medicine and Biology
PTB	Physikalisch Technische Bundesanstalt
RAM	Read Access Memory
ROI	Region of interest
RPI	Rensselaer Polytechnic Institute
SLAC	Stanford Linear Accelerator Center
SOP	Standard operating procedure
UBG	Underbust girth
UF	University of Florida
URL	Uniform Resource Locator
VTK	Visualization Toolkit
WBC	Whole Body Counter